PREFACE

Our intention in this mongraph is to survey a number of topics related to the study of the continuity and differentiation properties of real functions, in certain generalized senses. There is now a relatively large literature devoted to such subtle concerns but which is accessible and known only to specialists. Since the ideas are essentially simple and the techniques required fairly elementary this literature should be easily absorbed by any interested mathematician, and it is hoped that the presentation here is sufficiently readable and the exposition adequately clear for this purpose.

Probably the reader needs only a familiarity with the usual basics of real analysis (measure, category, density, etc.) in order to follow the arguments. This material is readily available in a variety of textbooks. A better preparation would be to master the books

S.Saks, Theory of the integral,

and

A.M.Bruckner, Differentiation of real functions,

(references [33] and [209] in the bibliography) that most analysts who work in this particular set of topics would surely consider fundamental to our subject. The present monograph continues certain concerns that arise in each of these works.

Part of this material was presented in a series of seminars at the University of California at Santa Barbara in the spring of 1984, during the special year in Real Analysis that was held there. I am particularly grateful and certainly indebted to the participants in that seminar who offered much helpful criticism and indicated numerous improvements. What remains is, doubtless, flawed but much less so than it would have been without the opportunity to meet with so many fine analysts.

In the first chapter is presented a general structure (called here a local system of sets) that can be used to formulate a variety of general notions of limit, continuity, derivative, etc. for real functions. The reason we have chosen this abstract framework is to enable us to clarify and codify the type of arguments that appear in the study. The greater generality itself is not of much interest; the real intention is to lay bare the underlying analysis. Thus it will appear that almost all of the arguments used in the subject reduce to a few general themes, most notably "intersection conditions" and various thickness conditions (porosity and density usually). All the basic concepts and arguments to be used in the rest of the work are introduced in the first chapter.

The second chapter is a review of the classical material on real cluster sets, from the perspective established in the first chapter. Again the basic arguments here, and elsewhere, will involve appropriate intersection conditions. This cluster set material is attractive and elementary, but does not seem to have been presented in any text to date leaving the interested reader to search through a large number of early references. By restricting ourselves to real cluster sets (i.e. cluster sets for functions of one real variable) we can present an apparently complete survey of the known results.

Chapter three contains a brief account of some general notions of continuity for real functions.

Chapter four gives an introduction to the notion of total variation for a real function. This presentation allows us to include some very classical material on functions of bounded variation, VBG* functions, singular functions, Lebesgue-Stieltjes measures, etc. from a perspective that is not well known and which allows a unified and simple treatment of some apparently diverse ideas.

In Chapter five we have given an account of several classes of monotonicity theorem. This should perhaps be read in conjunction with a study of Chapter XI of Bruckner's monograph ([33,pp.173-198]).

Chapters six and seven are devoted to a number of questions whose theme is the relationship that must hold among different types of generalized derivatives. This includes the well-known Denjoy-Young-Saks theorem and a variety of lesser known variants, both classical and recent.

Finally an Appendix is included that contains a survey on the notion of set porosity. This material too is not well known and can be found, so far, only scattered in the literature. As porosity computations and language appear in many instances in real analysis, this material should be of some use either as a point of reference or as an introduction to the concepts.

There are many more topics that could have been included and which would fit naturally within the framework that we are using. The properties of derivatives and extreme derivates in a generalized sense are currently being studied by some researchers. However this appears still to be in the early stages of development and we have chosen not to report on it. The interested reader should consult the article Bruckner, O'Malley and Thomson [43].

The bibilography contains many articles related to our concerns here, even if not explicitly discussed. It should not be considered complete, however, and the authors whose works I have not mentioned will forgive my oversight.

The notation used is mostly standard nowadays. Thus $A \cup B$, $A \cap B$, and $A \setminus B$ denote the usual union, intersection and difference of the sets A and B, while R denotes the set of real numbers and $\overline{A}$ the closure of the set A in R. However the peculiarities of the word processor have led to the somewhat old fashioned notation

$$\sum_{k=m}^{n} A_k \quad \text{and} \quad \prod_{k=m}^{n} A_k$$

for the union and intersection of a sequence of sets $\{A_k\}$. This notation should present no difficulties. Other special notations are explained in the text and may be found in the index.

Vancouver,B.C., 1985   B.S.T.

CONTENTS

CHAPTER ONE: LOCAL SYSTEMS

- §1. Introduction. 1
- §2. Local systems. 2
- §3. Partial order. 5
- §4. Dual systems. 7
- §5. Filters. 10
- §6. Limits of interval functions. 12
- §7. Derivates and derivatives. 14
- §8. Convergent interval functions. 16
- §9. Topological limits. 18
- §10. Systems based on a closure operator. 19
- §11. Negligent limits. 21
- §12. Density systems. 22
- §13. Miscellaneous limits. 24
- §14. Path approach. 27
- §15. Intersection conditions. 31
- §16. S-covers. 35
- §17. Variation of a function. 38
- §18. The variational measure. 40
- §19. The increasing sets property. 41

CHAPTER TWO: CLUSTER SETS

- §20. Introduction. 44
- §21. S-cluster sets. 44
- §22. Elementary theorems for cluster sets. 47
- §23. S-derived set. 50
- §24. Weak continuity theorems. 53
- §25. Asymmetry theorems. 56
- §26. Young's Rome theorem. 60
- §27. Ambiguity theorems. 62
- §28. Essential asymmetry. 64
- §29. Density cluster sets. 67

CHAPTER THREE: CONTINUITY

- §30. Introduction. 70
- §31. S-continuity. 70
- §32. S-discontinuities. 73
- §33. Baire 1 functions. 74
- §34. Negligent continuity. 78
- §35. Continuity relations. 80

CHAPTER FOUR: VARIATION OF A FUNCTION

§36. Introduction. 85
§37. Elementary properties. 86
§38. Variational estimates. 87
§39. Finite variation. 90
§40. σ-finite variation. 93
§41. The Vitali theorem. 96
§42. Derivates and variation. 98
§43. Further measure estimates. 100
§44. The Lebesgue-Denjoy-Lusin theorem. 103
§45. Absolute continuity. 105
§46. De la Vallée-Poussin decomposition theorem. 106
§47. Singular functions. 108

CHAPTER FIVE: MONOTONICITY

§48. Introduction. 115
§49. Local monotonicity. 115
§50. Generalized local monotonicity. 117
§51. Relative monotonicity. 120
§52. Further relative monotonicity theorems. 123
§53. Relative monotonicity (cont.). 129
§54. Global monotonicity theorems. 130
§55. Applications. 133
§56. The Goldowski-Tonelli theorem. 137

CHAPTER SIX: RELATIONS AMONG DERIVATES

§57. Introduction. 138
§58. Elementary relations. 139
§59. Beppo Levi theorem. 143
§60. Order relation for the negligent Dini derivates. 145
§61. A further order relation. 146
§62. Ward's theorems. 147
§63. Zajíček's relations. 149
§64. Theorem of W.H.Young. 151
§65. The porosity relations. 155
§66. Evans-Humke theorem. 158
§67. Denjoy/Khintchine/Burkill/HaslamJones theorem. 160
§68. Generalized Young-Evans-Humke theorem. 160
§69. Relations with the approximate derivative. 162

CHAPTER SEVEN: THE DENJOY-YOUNG RELATIONS

§70. The Denjoy-Young-Saks theorem. 166
§71. Proof of the DYS theorem. 167
§72. The approximate version. 170
§73. σ - porous version. 176
§74. The sharp version. 178
§75. DYS for the negligent Dini derivatives. 179

APPENDIX: SET POROSITY

§A1. Introduction. 183
§A2. Basic definitions. 183
§A3. Elementary properties. 187
§A4. σ-porous sets. 189
§A5. Porosity lemmas. 191
§A6. Zahorski's porosity derivative. 194
§A7. ψ-porosity. 197
§A8. σ-(ψ)-porosity. 200
§A9. Porosity and Hausdorff measure. 202
§A10. A further porosity lemma. 203
§A11. Symmetric perfect sets. 205

REFERENCES   213

LIST OF SPECIAL SYMBOLS   226

INDEX   227

CHAPTER ONE
LOCAL SYSTEMS

§1. Introduction.

The basic concepts of elementary analysis, (limits, continuity, derivatives, etc.) have undergone numerous generalizations in order to provide deeper insights into the structure of real functions. The beginning analysis student soon learns that all of the notions of "limit" that are introduced in a first course may be applied in a one sided version. Thus functions may have right or left hand limits, may be right or left continuous, and have right or left hand derivatives. Many authors have carried these ideas further by considering much more delicate refinements. The most important of the generalizations in this spirit has been the approximate limit and various related notions.

We shall give in this introductory section a definition of the approximate extreme limits that will help motivate the abstract structure that will be introduced in this chapter. Recall that the usual extreme limits of a function f at a point x_0, may be written in the following three equivalent formulations; each expresses essentially the same computation but suggests a different perspective on the limit operation. We may define $\limsup_{x \to x_0} f(x)$ as

$$\inf \{ y : \{ t : t = x_0 \text{ or } f(t) < y \} \text{ is a neighbourhood of } x_0 \},$$

or as

$$\inf \{ y : x_0 \text{ is not a limit point of } \{ t : t = x_0 \text{ or } f(t) > y \} \},$$

or, again equivalently,

$$\sup \{ y : x_0 \text{ is a limit point of } \{ t : t = x_0 \text{ or } f(t) > y \} \}.$$

Symmetrically we may define $\liminf_{x \to x_0} f(x)$ as

$$\sup \{ y : \{ t : t = x_0 \text{ or } f(t) > y \} \text{ is a neighbourhood of } x_0 \},$$

or

$$\sup \{ y : x_0 \text{ is not a limit point of } \{ t : t = x_0 \text{ or } f(t) < y \} \},$$

or

$$\inf \{ y : x_0 \text{ is a limit point of } \{ t : t = x_0 \text{ or } f(t) < y \} \}.$$

These expressions have led numerous authors to a number of generalizations of the notion of a limit for a real function. One of the earliest and most useful of these has been the concept of an approximate limit introduced by Denjoy (under that name) and by Khintchine (under the name "asymptotic limit"). The definitions above for the extreme limits use in the computations the idea that some set is either very thin at x_0 (does not have x_0 as a limit point) or is very fat at x_0 (is a full neighbourhood of x_0). The approximate limits are defined in a similar way but take thinness and thickness in the sense of density; thus the thinness is taken as density zero (that is upper, outer density zero) and the thickness is taken as full density 1 (that is lower, inner density 1).

For an arbitrary function f at a point x_0 one defines $\text{ap-lim sup}_{x \to x_0} f(x)$ as

$$\inf \{ y : \{ t : t = x_0 \text{ or } f(t) < y \} \text{ has (inner) density 1 at } x_0 \},$$

or

$$\inf \{ y : \{ t : t = x_0 \text{ or } f(t) > y \} \text{ has density zero at } x_0 \},$$

or

$$\sup \{ y : \{ t : t = x_0 \text{ or } f(t) > y \} \text{ has positive density at } x_0 \},$$

and with similar definitions for the lower approximate limit.

Once these definitions had been sufficiently studied it was natural that there would appear studies in which some analogous generalization would be used. Thus one can consider a number of alternative concepts that might might be used for the thinness and thickness notions here, in place of density.

With one sided versions, a spectrum of density type properties (upper, outer, inner, lower, various density values in $[0,1]$, etc.) and with further notions replacing density there has evolved a large literature devoted to the investigation of subtle properties of real functions within the language of certain of these generalized limits. The methods, even when they are similar, have not been systematically described, and the results often appear in a scattered fashion in the literature, and are frequently duplicated. It has become difficult to keep track of the results that have been obtained, the interrelations, and the general patterns.

In this chapter we outline a general structure which can be used to unify and simplify this study, and which will allow us to survey a broad range of results. The abstract notion of a "local system" will merely replace, formally, the above informal ideas relating to thickness and thinness. The duality between the two notions of thick and thin (expressed above in the observation that an extreme limit is defined as either an inf or a sup relative either to the thick or the thin notion) will be formalized too, and systematically used as a genuine dual notion.

The basic ideas are derived from many sources. The structure represents a mild generalization of the notion of limit used in topology, but adapted to the needs of certain problems in real analysis. Other authors have used similar structures. See, for example, Császár [52], Jędrzejewski [129], Tevy and Bruteanu [230], Świątkowski [226], and Zajíček [268].

§2. Local systems.

The framework that we shall use for our general notions of limit, continuity, derivation, etc. is a modification of the concept of a filter that is used in topology. Many authors have taken the definition of a filter and relaxed it in various ways, as for example in the notion of "sieve" or "quasi-filter". The setting we use is particularly convenient for expressing a large class of ideas in classical real analysis.

(2.1) DEFINITION. By a <u>local system</u> we mean a family S such that at each point $x \in \mathbb{R}$ there is given a nonempty collection of sets $S(x)$ with the following properties:
  (i) $\{x\} \notin S(x)$,
  (ii) if $S \in S(x)$ then $x \in S$,
  (iii) if $S_1 \in S(x)$ and $S_2 \supset S_2$ then $S_2 \in S(x)$,
  (iv) if $S \in S(x)$ and $\delta > 0$ then
  $$S \cap (x - \delta, x + \delta) \in S(x).$$

Such a system can be used for a variety of generalized notions and it is this notion that is to be exploited extensively throughout this work. In this section, by way of an introduction, we will present only some of the more basic ideas. If S is a local system then one can define a notion, relative to S, of a limit of a function f at a point x.

(2.2) DEFINITION. Let S be a local system, let f be a real function and let x be a point of $\mathbb{R}$. Then an (S)-<u>limit</u> of f at x is defined as any extended real number c,
$$(S)\text{-}\lim_{y \to x} f(y) = c$$
for which it is true that the set
$$\{t: t = x \text{ or } f(t) \in U_c\}$$
belongs to $S(x)$ for every neighbourhood U_c of c.

There is no requirement here that a limit be unique, and indeed in many applications it is the family of such numbers that is of interest. In such studies we shall denote the collection of all S-limits by the expression
$$(S)\text{-}\Lambda(f, x)$$
and refer to this collection by the term S-<u>cluster set</u>.

The <u>extreme limits</u> relative to a system S at a point x are defined as
$$(S)\text{-}\limsup_{y \to x} f(y) = \inf \{y: \{t: t = x \text{ or } f(t) < y\} \in S(x)\}$$
and
$$(S)\text{-}\liminf_{y \to x} f(y) = \sup \{y: \{t: t = x \text{ or } f(t) > y\} \in S(x)\}.$$

The example which follows will help make the intention of this definition clear as well as to exhibit the scope of its application. The two systems that are introduced in this example, S_0 and S_∞, will be used frequently in the sequel and we will use this notation throughout our study.

(2.3) Example. Let S_0 denote the system defined at each point x as

$$S_0(x) = \{ S : S \text{ contains an open interval about the point } x \}$$

so that each $S_0(x)$ is precisely the neighbourhood filter at the point x.

We define a closely associated system S_∞ defined at each point x as

$$S_\infty(x) = \{ S : S \text{ contains } x \text{ and has } x \text{ as an accumulation point} \}.$$

The limits defined above then can be easily seen to have the following properties:

$$(S_0)\text{-}\lim_{y \to x} f(y) = \lim_{y \to x} f(y)$$

where the limit is taken in the usual sense; and (more dramatically)

$$(S_\infty)\text{-}\lim_{y \to x} f(y) = c$$

if and only if there is at least one sequence $x_n \to x$ ($x_n \neq x$) such that the sequence $f(x_n)$ converges to c. In particular it is clear that the limits in the S_∞ sense are normally not unique.

For the extreme limits relative to these two systems we have a remarkable property: these extreme limits are again just the usual extreme limits but with an interesting reversal for the system S_∞.

One has

$$(S_0)\text{-}\limsup_{y \to x} f(y) = (S_\infty)\text{-}\liminf_{y \to x} f(y) = \limsup_{y \to x} f(y),$$

and

$$(S_0)\text{-}\liminf_{y \to x} f(y) = (S_\infty)\text{-}\limsup_{y \to x} f(y) = \liminf_{y \to x} f(y).$$

This feature of the two systems offers us a duality for local systems that we will exploit in section 4. The unusual feature of these limits can be made more intuitive by reviewing the material in the introduction. It is an elementary fact that an extreme limit may be viewed as a sup-inf or equally well as an inf-sup.

The systems S_0 and S_∞ do not merely serve as illustrations of the theory, but actually lie at the two extremes permitted by the definition. We may express this in an easily proved lemma.

(2.4) LEMMA. For any local system S, and at any point x, one has invariably the set inclusions

$$S_0(x) \subset S(x) \subset S_\infty(x).$$

PROOF. By the way that a local system S has been defined it is clear that, at any point x, each neighbourhood $(x-\delta, x+\delta)$ must belong to $S(x)$ and that each set $S \in S(x)$ must have x as a limit point. This is precisely what the two set inclusions say.

§3. Partial order.

It is natural in this setting to introduce a partial order on the family of local systems, so that we have a lattice structure available. Although we have no great need of an elaborate formal treatment of such structures, it is nonetheless convenient to introduce this language and to exploit it to some degree in the theory.

(3.1) DEFINITION. Let S_1 and S_2 be two local systems. We will write $S_1 \ll S_2$ if, at every point x, there is the inclusion $S_1(x) \subset S_2(x)$.

It is clear that this relation is a partial order on the family of local systems that allows us to rewrite lemma (2.4) above as asserting the relation
$$S_0 \ll S \ll S_\infty.$$
Thus there are two extremes in the partial order. In fact the partial order has the structure of a lattice with the following definitions of the lattice operations.

(3.2) DEFINITION. Let S_1 and S_2 be two local systems. Then we define the systems $S_1 \vee S_2$ and $S_1 \wedge S_2$ by writing at each point x,
$$[S_1 \vee S_2](x) = S_1(x) \cup S_2(x),$$
and
$$(S_1 \wedge S_2](x) = S_1(x) \cap S_2(x).$$
It is easy to check that the terms $S_1 \vee S_2$ and $S_1 \wedge S_2$ are themselves local systems, and that one has the partial order relations
$$S_1 \wedge S_2 \ll S_i \ll S_1 \vee S_2 \qquad (i = 1, 2).$$

(3.3) Example. Let S_0 denote the usual neighbourhood system. We can define right and left versions of this by writing
$$S_0^+(x) = \{ U : U \text{ a right neighbourhood of } x \}$$
and
$$S_0^-(x) = \{ U : U \text{ a left neighbourhood of } x \}.$$
Then the systems S_0, S_0^+, and S_0^- are related by the assertion
$$S_0 = S_0^+ \wedge S_0^-.$$

Interpreted in terms of limits these lattice operations permit the following expressions.

(3.4) LEMMA. Let S_1 and S_2 be local systems and let f be an arbitrary function. Then one has the assertion

$$(S_1 \wedge S_2)\text{-}\lim_{y \to x} f(y) = c$$

if and only if both of the assertions

$$(S_1)\text{-}\lim_{y \to x} f(y) = c \quad \text{and} \quad (S_2)\text{-}\lim_{y \to x} f(y) = c$$

are valid. Similarly one has the assertion

$$(S_1 \vee S_2)\text{-}\lim_{y \to x} f(y) = c$$

if and only if one at least of the assertions

$$(S_1)\text{-}\lim_{y \to x} f(y) = c \quad \text{or} \quad (S_2)\text{-}\lim_{y \to x} f(y) = c$$

is valid.

PROOF. This is straightforward.

The relations that are expressed in this lemma may be equally well asserted in the language of cluster sets.

(3.5) COROLLARY. Let S_1 and S_2 be local systems and let f be an arbitrary function. Then the cluster sets of f relative to the systems S_1, S_2, $(S_1 \vee S_2)$, and $(S_1 \wedge S_2)$ have the following relations

$$(S_1 \vee S_2)\text{-}\Lambda(f, x) = (S_1)\text{-}\Lambda(f, x) \cup (S_2)\text{-}\Lambda(f, x)$$

and

$$(S_1 \wedge S_2)\text{-}\Lambda(f, x) = (S_1)\text{-}\Lambda(f, x) \cap (S_2)\text{-}\Lambda(f, x).$$

(3.6) Example. Lemma (3.4) and its corollary assert an elementary fact about limits which the abstract language may obscure. Let S_0 denote the usual neighbourhood system and let S_0^+ and S_0^- denote the right and left neighbourhood systems. Then $S_0 = S_0^+ \wedge S_0^-$ and the lemma just gives a general abstract expression of the fact that a limit $\lim_{y \to x} f(y)$ can exist if and only if both the right hand and the left hand limits

$$\lim_{y \to x+} f(y) \quad \text{and} \quad \lim_{y \to x-} f(y)$$

exist and have the same value. Similarly, using S_∞, S_∞^+, and S_∞^-, a number is a cluster set value at a point if and only if it is a cluster value on either the right or the left at that point.

Parallel to the assertion in (3.4) we have the following similar assertion for the extreme limits.

(3.7) LEMMA. Let S_1 and S_2 be local systems and let f be an arbitrary function. Then
$$(S_1 \wedge S_2)\text{-lim sup}_{y \to x} f(y)$$
is the maximum
$$\max \{ (S_1)\text{-lim sup}_{y \to x} f(y), (S_2)\text{-lim sup}_{y \to x} f(y) \},$$
and
$$(S_1 \vee S_2)\text{-lim sup}_{y \to x} f(y)$$
is the minimum
$$\min \{ (S_1)\text{-lim sup}_{y \to x} f(y), (S_2)\text{-lim sup}_{y \to x} f(y) \}.$$

§4. Dual systems.

The notion of a dual system is a most convenient tool in this theory. We have seen, in example (2.3), the parallel role played by the two systems S_0 and S_∞. Any concept defined in terms of one may be equally well redefined, in some way, in terms of the other. The key property that interconnects these two systems can be expressed by the fact that a set S will have
$$S \cup \{x\} \in S_0(x)$$
if and only if
$$(\mathbb{R} \setminus S) \cup \{x\} \notin S_\infty(x).$$
Conversely (and symmetrically)
$$S \cup \{x\} \in S_\infty(x)$$
if and only if
$$(\mathbb{R} \setminus S) \cup \{x\} \notin S_0(x).$$
Loosely put this says only that a set is a neighbourhood of a point x if and only if the complement of that set fails to have that point as a point of accumulation. The point x itself, of course, must be included in the sets of our system and so the complementation idea must be written in the above way.

This notion generalizes easily, so that to any local system corresponds a "dual" system defined in precisely this manner.

(4.1) DEFINITION. Let S be a local system. Then by the <u>dual</u> of S, denoted S^*, we mean the system defined so that for $x \in \mathbb{R}$, $S \in S^*(x)$ if and only if

(i) $x \in S$, and

(ii) $(\mathbb{R} \setminus S) \cup \{x\} \notin S(x)$.

The following facts about the dual of a local system follow directly from the definition.

(4.2) **LEMMA.** If S is a local system then so too is the dual family S^*.

PROOF. We have only to check that each of the requirements of definition (4.1) is satisfied. Firstly it is clear that, at any point x, the set $\{x\}$ cannot belong to S^*; to see this we observe that if $\{x\}$ belongs to S^* then $(\mathbb{R} \setminus \{x\}) \cup \{x\} = \mathbb{R}$ cannot belong to $S(x)$ which is impossible. The second property, that for each $S \in S^*(x)$, the point x must be in S is immediate.

Suppose now that $S_1 \supset S_2 \in S^*(x)$. Then by definition
$$(\mathbb{R} \setminus S_2) \cup \{x\}$$
cannot be in $S(x)$. But
$$(\mathbb{R} \setminus S_1) \cup \{x\} \subset (\mathbb{R} \setminus S_2) \cup \{x\}$$
so that
$$(\mathbb{R} \setminus S_1) \cup \{x\}$$
cannot be in $S(x)$ either, thereby proving that $S_1 \in S^*(x)$ as required.

Finally for any $S \in S^*(x)$ and any $\delta > 0$ we need to show that the set
$$S \cap (x - \delta, x + \delta)$$
also belongs to $S^*(x)$. If this set does not belong, then
$$(\mathbb{R} \setminus (S \cap (x - \delta, x + \delta))) \cup \{x\}$$
must be in $S(x)$. By the properties of S this requires that
$$(\mathbb{R} \setminus S) \cup \{x\} \in S(x),$$
and this contradicts the fact that S is in S^*. Thus we have verified each of the requirements, and it follows that the system S^* must be a local system.

(4.3) **LEMMA.** The systems S_0 and S_∞ are dual, i.e. $S_0^* = S_\infty$ and $S_\infty^* = S_0$.

(4.4) **LEMMA.** Let S be a local system. Then S^* is also a local system and its dual S^{**} is identical to S.

PROOF. We have already seen that S^* must be a local system. It is an easy matter to prove, directly from the definition, that $S \in (S^*)^*(x)$ if and only if $S \in S(x)$.

The feature of the dual S_∞ of the system S_0, given in example (2.3), whereby it served to reverse the lim inf and the lim sup is in general available for any dual pair.

(4.5) **LEMMA.** Let S be a local system and let S^* be its dual. Then for any function f the extreme limits must satisfy

$$(S)\text{-}\limsup_{y \to x} f(y) = (S^*)\text{-}\liminf_{y \to x} f(y),$$

and

$$(S)\text{-}\liminf_{y \to x} f(y) = (S^*)\text{-}\limsup_{y \to x} f(y),$$

at every point x.

PROOF. Suppose that $c = (S)\text{-}\limsup f(x)$ which, to simplify the arguments, we will consider as finite. Let $\delta > 0$ be given. Then consider the two sets

$$S_1 = \{y : y = x \text{ or } f(y) < c + \delta\}$$

and

$$S_2 = \{y : y = x \text{ or } f(y) < c - \delta\}.$$

We know that $S_1 \in S(x)$ and that $S_2 \notin S(x)$. Passing to the dual S^*, we obtain

$$(\mathbb{R} \setminus S_1) \cup \{x\} \notin S^*$$

and

$$(\mathbb{R} \setminus S_2) \cup \{x\} \in S^*.$$

Translating this, we obtain

$$\{y : y = x \text{ or } f(y) \geq c + \delta\} \notin S^*$$

and

$$\{y : y = x \text{ or } f(y) \geq c - \delta\} \in S^*.$$

From this we see, by definition, that

$$c - \delta < (S^*)\text{-}\liminf_{y \to x} f(y) < c + \delta$$

and, since $\delta > 0$ is arbitrary, we obtain that $(S)\text{-}\limsup f(x)$ and $(S^*)\text{-}\liminf f(x)$ are identical as required.

The relation between the dual and the partial order is expressed in the next lemma. Note that the dual operation reverses the order.

(4.6) **LEMMA.** Let S_1 and S_2 be local systems. Then the following must hold:
  (a) if $S_1 \ll S_2$ then $S_2^* \ll S_1^*$.
  (b) $(S_1 \vee S_2)^* = S_1^* \wedge S_2^*$.
  (c) $(S_1 \wedge S_2)^* = S_1^* \vee S_2^*$.

PROOF. Each of these requires just a detailed application of the definitions in a routine manner. For example if S is an element of $S_2^*(x)$ then, to prove (a), we must show that S also belongs to $S_1^*(x)$. This means that $(\mathbb{R} \setminus S) \cup \{x\}$ does not belong to $S_2(x)$. By the order relation $S_1 \ll S_2$, we know then that $(\mathbb{R} \setminus S) \cup \{x\}$ does not belong to $S_1(x)$. Consequently S must be in S_1^* as required.

§5. Filters.

In a great many cases the system S that satisfies the four conditions of definition (2.1) has an extra property that is of considerable use. We will say that S is <u>filtering</u> at a point x if it the case that $S_1 \cap S_2 \in S(x)$ whenever S_1 and S_2 belong to $S(x)$.

It is clear that if a local system S is filtering at x then the family of sets $S(x)$ is a filter converging to x. Conversely, if at each point x there is given a filter $S(x)$ converging to x and nontrivial at x in the sense that $\{x\}$ does not belong to $S(x)$, then S is a local system.

The system S_0, in example (2.3), is at each x a filter converging to x. This is not the case for the system S_∞ since two sets S_1 and S_2 may belong to $S_\infty(x)$ and yet the intersection $S_1 \cap S_2$ may contain no more than the point $\{x\}$.

In this section we begin by showing that the fact of a system being filtering will require some property of the dual. This is a useful fact in that when a system S has a dual S^* that is filtering there are a number of useful consequences that can be easily proved from the characterization available here. This is the first of what we shall consider as <u>dual properties</u>; if a system S has some property then its dual S^* must inherit some parallel property. For each property that we introduce on a system we should properly consider an expression of its dual property. In the lemma below we give precisely the property that is dual to the filtering property. For convenience of reference we shall denote this property as $[J_1]$, after Jędrzejewski [129], where it plays a key role.

(5.1) LEMMA. Let S be a local system. Then its dual S^* is filtering at a point x if and only if the system S has the following property ($[J_1]$): if S_1 and S_2 both contain the point x and $S_1 \cup S_2 \in S(x)$ then necessarily either $S_1 \in S(x)$ or $S_2 \in S(x)$.

PROOF. Suppose S^* is filtering at x. We shall show that whenever S_1 and S_2 are not in $S(x)$ but both contain x, then the union $S_1 \cup S_2$ cannot be in $S(x)$. This will prove the lemma in one direction. If S_1 and S_2 do not belong to $S(x)$ then by definition $(\mathbb{R} \setminus S_1) \cup \{x\}$ and $(\mathbb{R} \setminus S_2) \cup \{x\}$ must both belong to $S^*(x)$. As S^* is supposed to be filtering at x it follows that the intersection

$$\{(\mathbb{R} \setminus S_1) \cap (\mathbb{R} \setminus S_2)\} \cup \{x\}$$

is also in $S^*(x)$. But this means that $S_1 \cup S_2$ cannot belong in the dual of S^*, that is

$$S_1 \cup S_2 \notin S^{**}(x) = S(x)$$

exactly as we wished to prove. A reversal of these arguments will prove the lemma in the other direction.

(5.2) LEMMA. Let S be a local system that is filtering at a point x. Then at x there must be the inclusion $S^*(x) \supset S(x)$.

PROOF. If $S \in S(x)$ and $S(x)$ is filtering then we can show that $S \in S^*(x)$. If not then $S \notin S^*(x)$ and by definition this means that $(\mathbb{R} \setminus S) \cup \{x\}$ must be in $S^{**}(x) = S(x)$. But this is impossible since then the intersection
$$[S \cap (\mathbb{R} \setminus S)] \cup \{x\} = \{x\}$$
must belong to $S(x)$. But, by definition, this set must not belong. From this contradiction the proof is obtained.

(5.3) COROLLARY. Let S be filtering. Then $S \ll S^*$.

These features of a filtering system can be proved with a weaker assumption. In the above proof we took a pair of sets S_1 and S_2 from $S(x)$ and we required that the intersection $S_1 \cap S_2$ contain more than the single point x; the filtering assumption certainly provides this, but less may be assumed. Thus we have a further corollary.

(5.4) COROLLARY. Let S be a local system that has the property that, for any point x and for any sets S_1 and S_2 in the collection $S(x)$, the intersection $S_1 \cap S_2$ contains more than the point x. Then $S \ll S^*$.

It is convenient to place here one of the main features of limits taken with respect to systems that are filtering, or, more generally, which satisfy the order relation expressed above.

(5.5) LEMMA. Let S be a system that satisfies at a point x the requirement that
$$S_1 \cap S_2 \neq \{x\}$$
for every pair of sets S_1 and S_2 from $S(x)$. Then for any function f one has at x the inequality,
$$(S)\text{-}\liminf_{y \to x} f(y) \leq (S)\text{-}\limsup_{y \to x} f(y)$$
and if $(S)\text{-}\lim_{y \to x} f(y) = c$ then c is unique.

PROOF. Since we have $S(x) \subset S^*(x)$ we know that
$$(S)\text{-}\liminf_{y \to x} f(y) = (S^*)\text{-}\limsup_{y \to x} f(y)$$
$$\leq (S)\text{-}\limsup_{y \to x} f(y)$$
which proves the inequality. The uniqueness of the limits follows similarly.

(5.6) Remark. The subject of ultrafilters might naturally be introduced here. In fact there is an intimate connection between ultrafilters and systems that are self dual, i.e.

systems S for which $S = S^*$. However we shall not require any such material in the sequel and so leave this topic for the reader to pursue elsewhere.

§6. Limits of interval functions.

In addition to studying various generalized limits for real functions relative to a local system we need as well a notion of limit for interval functions. In particular we shall be interested in studying generalized derivatives and this requires a study of the interval function

$$I \to \frac{\Delta f(I)}{|I|}$$

(where $I = [x, y]$, $|I| = y - x$, and $\Delta f(I) = f(y) - f(x)$).

The definition we give assumes that S is a local system and that ψ is a real valued interval function defined for every interval I. Refinements can be made on this for certain purposes. We will always consider that the interval $[a, b]$ for $a < b$ can also be written as $[b, a]$; this allows for more economical assertions in many proofs.

(6.1) DEFINITION. For any local system S and any interval function ψ we define an (S)-limit

$$(S)\text{-lim}_{I \to x} \psi(I) = c$$

for an extended real number c, if the set of points

$$\{ t : t = x \text{ or } \psi([t, x]) \in U_c \}$$

belongs to $S(x)$ for every neighbourhood U_c of c. The extreme S-limits of ψ are defined as

$$(S)\text{-limsup}_{I \to x} \psi(I) = \inf{}_{S \in S(x)} \sup\{ \psi([x, y]) : y \in S, y \neq x \},$$

and

$$(S)\text{-liminf}_{I \to x} \psi(I) = \sup{}_{S \in S(x)} \inf\{ \psi([x, y]) : y \in S, y \neq x \}.$$

Note that each of these limits is really an endpoint limit. The values of the interval function ψ at an interval $[a, b]$ with $a < x < b$ do not affect the computation of the limit at the point x. In some applications it is necessary to use a more restrictive version of an interval function limit. We shall call this a strict limit, meant to indicate that further values of the interval function are being considered. (The "s" in the limit operation might also be interpreted to mean "strong" or "straddled".)

(6.2) DEFINITION. Let ψ be an interval function and S a local system. Then a __strict limit__ of ψ at a point x relative to S is defined as
$$(S)\text{-slim}_{I \to x}\psi(I) = c$$
provided for every neighbourhood U_c of c there is a set $S \in S(x)$ such that whenever $y \leq x \leq z$, $y, z \in S$, $y \neq z$, one has
$$\psi([y, z]) \in U_c \,.$$

Similarly the __strict extreme limits__ are defined as
$$(S)\text{-slim sup}_{I \to x}\psi(I) = \inf{}_{S \in S(x)} \sup\{\psi([y, z]) : y, z \in S, y \leq x \leq z, y \neq z\}$$
and
$$(S)\text{-slim inf}_{I \to x}\psi(I) = \sup{}_{S \in S(x)} \inf\{\psi([y, z]) : y, z \in S, y \leq x \leq z, y \neq z\} \,.$$

For interval functions of the special form $\psi(I) = \Delta f(I)/\Delta g(I)$, such as would appear in any computations involving derivatives or relative derivatives, the strict limits are identical with the simpler limits. We give this in a lemma.

(6.3) LEMMA. Let S be a local system and let ψ be an interval function that is of the special form
$$\psi(I) = \frac{\Delta f(I)}{\Delta g(I)}$$
for two real functions f and g. Then the S-limits and the S-strict limits are identical.

PROOF. Suppose that $y \leq x \leq z$ and that
$$\left| \frac{\Delta f([y, x])}{\Delta g([y, x])} - c \right| < \delta$$
and
$$\left| \frac{\Delta f([x, z])}{\Delta g([x, z])} - c \right| < \delta$$
for all y, z ($y \neq z$) in a set S. Then by the additivity of the two interval functions Δf and Δg one obtains easily that
$$\left| \frac{\Delta f([y, z])}{\Delta g([y, z])} - c \right| < \delta$$
for all such y, z and the lemma can be proved. Here we have assumed that the function ψ is well defined at least locally so that we do not need to worry about a zero denominator.

§7. Derivates and derivatives.

For later reference we will fix the notation to be used for derivates relative to a local system. These notions will be studied in much greater detail in Chapters 4, 5, 6, and 7.

(7.1) DEFINITION. Let f be a real function and S a local system. Then the extreme S-limits of the interval function

$$I \to \frac{\Delta f(I)}{|I|}$$

at a point x are denoted as

$$(S)\text{-}\overline{D} f(x) \quad \text{and} \quad (S)\text{-}\underline{D} f(x)$$

and referred to as the <u>extreme derivates</u> of f at x relative to the system S. If an S-limit exists then it is written

$$(S)\text{-}D f(x)$$

and called a <u>derivative</u> of f at x relative to S.

(7.2) Example. Consider the S_0 and S_∞ limits of an interval function ψ of the form

$$\psi(I) = \frac{\Delta f(I)}{|I|}$$

where f is a real function, Δf its associated interval function, and $|I|$ denotes the length of the interval I. Then the limit of the interval function ψ with respect to the system S_0 is just

$$(S_0)\text{-}\lim \psi(x) = \lim_{y \to x} \frac{f(y) - f(x)}{y - x}$$

and so it is the ordinary derivative $f'(x)$ if it exists. For an S_∞-limit of this same interval function we have that

$$(S_\infty)\text{-}\lim \psi(x) = c$$

if and only if there is at least one sequence of numbers $\{x_n\}$ different from x and converging to x so that

$$\lim_{n \to +\infty} \frac{f(x_n) - f(x)}{x_n - x} = c.$$

Consequently any derived number of the function f at x is an S_∞-limit of the interval function ψ. The extreme limits of this interval function have an obvious relation with the extreme derivates of f.

We will use the following notation for special derivates taken relative to certain systems that arise frequently in our study.

(7.3) For the <u>extreme bilateral derivates</u>, taken with respect to the system S_0 we write,

$$\overline{D} f(x) \quad \text{and} \quad \underline{D} f(x).$$

(7.4) For the <u>extreme unilateral derivates</u>, better known perhaps as the "Dini derivatives", taken relative to the systems S_o^+ and S_o^- we write,
$$\overline{D}^+ f(x) \text{ and } \underline{D}^+ f(x),$$
and
$$\overline{D}^- f(x) \text{ and } \underline{D}^- f(x).$$

(7.5) For the <u>extreme bilateral approximate derivates</u> (as will be defined in section 12 below) we adopt the notation,
$$\underline{D}_{ap} f(x) \text{ and } \overline{D}_{ap} f(x),$$
and for the one-sided versions of these (also known as the "approximate Dini derivatives") we write
$$\underline{D}_{ap}^+ f(x) \text{ and } \overline{D}_{ap}^+ f(x),$$
and
$$\underline{D}_{ap}^- f(x) \text{ and } \overline{D}_{ap}^- f(x).$$

(7.6) For the <u>extreme derivates taken by neglecting sets in a given ideal of sets</u> N (a notion that is defined in section 11 below) we write,
$$\underline{D}_N f(x) \text{ and } \overline{D}_N f(x),$$
with a similar notation for the one-sided versions:
$$\underline{D}_N^+ f(x) \text{ and } \overline{D}_N^+ f(x),$$
and
$$\underline{D}_N^- f(x) \text{ and } \overline{D}_N^- f(x).$$

(7.7) We shall also need a notation for the "sharp" derivates defined as
$$\overline{D}^\# f(x) = \limsup\nolimits_{y, z \to x, y \neq z} \frac{f(y) - f(z)}{y - z}$$
and
$$\underline{D}^\# f(x) = \liminf\nolimits_{y, z \to x, y \neq z} \frac{f(y) - f(z)}{y - z}.$$
Note that these derivates do not fit our pattern and are not expressible as system limits.

Many more types of derivations will be investigated in later chapters but, to keep the notation to a minimum, no further standardized terminology will be enforced.

§8. Convergent interval functions.

We continue our study of interval function limits with an observation on the nature of the limits of an S_o-convergent interval function. This is related to a similar theorem of Gleyzal [98] but his theorem does not require of the interval function that it be everywhere defined. The theorem can be traced back to Baire [4]. While the techniques here are classical this presentation will help shape the tools that are of most use to us in this study. For a discussion of the origin of Baire's ideas and an account of their later development see the article of Laczkovich [143]; for a survey of the results on Baire 1 functions that are also Darboux see the article of Ceder and Pearson [48].

(8.1) THEOREM. A real function f is in the first class of Baire if and only if it can be expressed as the S_o-limit of a convergent interval function ψ,

$$(S_o)\text{-}\lim_{I \to x} \psi(I) = f(x)$$

for every $x \in \mathbb{R}$.

PROOF. Suppose firstly that f is expressible as such a limit but that, contrary to the theorem, f is not Baire 1. Then by a common device (cf. Bruckner [33, p.104]) there must be a perfect set Q and a positive number δ such that the oscillation of f on Q exceeds δ at every point of Q. Choose for each $x \in Q$ a positive number $\nu(x)$ so that

$$|\psi([x,y]) - f(x)| < \frac{\delta}{6}$$

for every number y with $y \neq x$ and $|y - x| < \nu(x)$.

We use the positive function ν to induce a partition of the set Q as follows: write for each $n = 1, 2, 3, \ldots$, and $m = 0, \pm 1, \pm 2, \ldots$

$$Q_{nm} = \{ x \in Q : x \in [\tfrac{m}{n}, \tfrac{m+1}{n}),\ \tfrac{1}{n} < \nu(x) \leq \tfrac{1}{n-1} \}.$$

One of these sets Q_{nm} in the partition must be dense in a portion of the set Q (this by Baire's theorem). If a set Q_{nm} is dense in $Q \cap (c,d)$ then we shall arrive at a contradiction. For if z_1, z_2 are any two points in this portion of Q there must be points x and y in Q_{nm} with the property that

$$|z_1 - x| < \min\{\nu(z_1), \nu(x)\},$$
$$|x - y| < \min\{\nu(x), \nu(y)\},$$

and

$$|y - z_2| < \min\{\nu(y), \nu(z_2)\}.$$

From this we get the inequalities

$$|\psi([z_1, x]) - f(z_1)| < \tfrac{\delta}{6}, \quad |\psi([z_1, x]) - f(x)| < \tfrac{\delta}{6}$$
$$|\psi([x, y]) - f(x)| < \tfrac{\delta}{6}, \quad |\psi([x, y]) - f(y)| < \tfrac{\delta}{6}$$

$$\left|\psi([z_2,y])-f(y)\right| < \frac{\delta}{6}, \text{ and } \left|\psi([z_2,y])-f(z_2)\right| < \frac{\delta}{6}$$

so that, in the obvious fashion, one concludes that

$$\left|f(z_1)-f(z_2)\right| < \delta.$$

Since this is to hold for every pair of numbers z_1, z_2 in this portion this contradicts our statement about f and Q, thereby proving the first part of the theorem.

In the other direction let us suppose that f is in the first class of Baire. We must show how to construct an interval function ψ so that f is the S_0-limit of this interval function at every point. Since f is Baire 1 there must be a sequence of continuous functions $\{f_n\}$ converging pointwise to f. We simplify our construction by placing everything within the unit interval $[0,1]$ but the construction can be extended to the rest of the line. Set $f_0 = 0$, $\kappa_0 = 1$ and choose numbers $\{\kappa_n\}$ so that

$$0 < \kappa_n < \kappa_{n-1},$$

and

$$\left|f_n(x)-f_n(y)\right| < \frac{1}{n}$$

when $x, y \in [0,1]$, $|x-y| < \kappa_n$ (this just uses the uniform continuity of f_n on the interval $[0,1]$). For any interval $[a,b] \subset [0,1]$ we define the interval function ψ to assume the value

$$\psi([a,b]) = f_n(a)$$

where the n is chosen so that $\kappa_{n+1} < b-a \leq \kappa_n$.

This then defines ψ properly on every subinterval of $[0,1]$ and it remains only to show that $(S_0)\text{-}\lim \psi(x) = f(x)$ at each point.

Suppose that $x \in [0,1]$ and $\nu > 0$ are given. Then we may choose an integer m so that

$$\frac{1}{m} < \frac{\nu}{2}$$

and so that

$$\left|f_n(x)-f(x)\right| < \frac{\nu}{2}$$

for all $n \geq m$. Then if an interval $[x,y]$ has $0 < |y-x| < \kappa_m$ we know that there must be an integer $n \geq m$ so that $\psi([x,y]) = f_n(x)$ (for $x < y$) or that $\psi([x,y]) = f_n(y)$ (for $y < x$) and such that

$$\left|f_n(x)-f_n(y)\right| < \frac{1}{n} < \frac{1}{m} < \frac{\nu}{2}.$$

Here if $x < y$ then we have immediately that

$$\left|\psi([x,y])-f(x)\right| = \left|f_n(x)-f(x)\right| < \nu.$$

If $y < x$ then

$$\left|\psi([x,y])-f(x)\right| \leq \left|f_n(y)-f_n(x)\right| + \left|f_n(x)-f(x)\right| < \nu$$

so that in either case we have the correct inequality to verify that f is everywhere in $[0,1]$ the limit of the interval function ψ.

§9. Topological limits.

In this and several subsequent sections we collect a variety of examples that ilustrate the general theory and which we will call upon for special results. These examples should help to reveal the motivations behind much of the theory, as well as to provide concrete situations in which the general and abstract notions may be realized. Indeed it is this collection of examples and the multiplicity of results and methods that they have generated that has prompted this abstract study; it is not the abstract study itself that is of great interest to us but merely the fact that this general structure permits a systematic account of the well known examples presented here, helps to reveal more clearly why it is that certain properties are shared, and clarifies the reasons that certain properties are not shared.

We have already defined above, in section 2, the system S_0 and its dual S_∞. These have been described as:

$$S_0(x) = \{ S : S \text{ any neighbourhood of } x \},$$

and

$$S_\infty(x) = \{ S : x \in S \text{ and } x \text{ is a limit point of } S \}.$$

We also define one-sided verions of these two systems by considering only right (left) neighbourhoods and right (left) points of accumulation. These we will denote as S_0^+, S_0^-, S_∞^+, and S_∞^-.

Recall that we have established the following duality relations among these systems:

$$S_0^* = S_\infty, \quad S_\infty^* = S_0$$
$$(S_0^+)^* = S_\infty^+, \text{ and } (S_\infty^+)^* = S_0^+.$$

Each of the systems S_0, S_0^+, and S_0^- expresses limits that arise from topologies on the real line. In a similar way we may let τ denote any topology on the real line that is finer than the usual real topology and for which no singleton set $\{x\}$ is open in τ. Then if we take

$$S_\tau(x) = \{ S : S \text{ is a } \tau\text{-neighbourhood of } x \},$$

and for the dual

$$S_\tau^*(x) = \{ S : x \in S \text{ and } x \text{ is a } \tau\text{-accumulation point of } S \},$$

the systems S_τ and S_τ^* are dual local systems that directly generalize the systems above. The (S_τ)- and (S_τ^*)-limits have an obvious topological interpretation.

It is a result of elementary topology that a topology may be characterized by its neighbourhood system. Thus the system characterizes the topology and any topology generates a neighbourhood system. The connection is given in the following lemma, whose proof may be found in most elementary topology texts.

(9.1) LEMMA. Let S be a local system. Then S is the system S_τ generated by a topology τ if and only if S is filtering and has the following property: for any point x and any set $S \in S(x)$ there is a set $S_0 \in S(x)$ with $S_0 \subset S$ and such that $S_0 \in S(y)$ for every point $y \in S_0$.

PROOF. See, for example, Willard [253, Theorem 4.2, p.31].

(9.2) Example. (The density topology). The most interesting example of a topological limit on the real line in which the topology is not immediately presented, is provided by the approximate limit. This will be discussed in greater detail below in section 12; here we mention only that the collection of sets τ_d, defined as including every measurable set A with density 1 at each of its points, is a topology on IR called usually "the density topology". Limits taken in the sense of this topology are the familiar approximate limits.

For details on the density topology see Goffman and Watermann [104], Goffman, Neugebauer and Nishiura [103], Oxtoby [194], Tall [228], and Ostaszewski [192], [193]. This topology has been shown to be completely regular, but not normal, not first countable, and not Lindelöf.

§10. Systems based on a closure operator.

The topological formulation of the preceeding section can be relaxed in certain settings by introducing what is called a <u>closure operator</u>. By a closure operator u we will understand a mapping
$$u : 2^{IR} \to 2^{IR}$$
with the following properties:
(i) $u(M) \supset M$ for every set M,
(ii) u satisfies a finite union property, namely that
$$u\left(\sum_{i=1}^{m} M_i\right) = \sum_{i=1}^{m} u(M_i)$$
for any finite sequence of sets $\{M_1, M_2, M_3, \ldots, M_m\}$, and
(iii) u is finer than the usual closure operator on the real line in the sense that for any set A the set $u(A)$ must be a subset of the points of accumulation of A.

Relative to such a closure operator we define the <u>u-derivative</u> of a set X to be
$$\text{der}_u(X) = \{t : t \in u(X \setminus \{t\})\}.$$
Then we may define a local system by the following device. Let u be a closure operator as defined above and denote by S_u the system
$$S_u(x) = \{S : x \in \text{der}_u(S \setminus \{x\})\}.$$
We claim that S_u is a local system. Indeed any system S that has a filtering dual may

be described by such a closure operator. We express these facts in a lemma.

(10.1) LEMMA. Let u be a closure operator. Then S_u is a local system with a filtering dual. Conversely, if S is a local system with a filtering dual, then there is a closure operator u so that $S = S_u$.

PROOF. Let us firstly check that the system S_u as defined above satisfies each of the requirements of a local system. Certainly at any point x the set $\{x\}$ cannot belong to $S_u(x)$ since $u(\emptyset) = \emptyset$. By definition if $S \in S_u(x)$ then $x \in S$.

Suppose now that $S_1 \supset S_2$. Then because of the additivity of u the set $u(S_1)$ is a subset of $u(S_2)$. Thus one checks easily that, whenever S_2 belongs to a collection $S_u(x)$, so too must any set S_1 that contains S_2. Finally, to check the remaining property, suppose that $S \in S_u(x)$ is given and that δ is a positive number. Consider the sets
$$S_1 = S \setminus \{x\},$$
and
$$S_2 = S \cap (x - \delta, x + \delta) \setminus \{x\}.$$
We have that
$$u(S_1) = u(S_2) + u(S_1 \setminus S_2).$$
But the point x cannot belong to $u(S_1 \setminus S_2)$ since it cannot be an ordinary point of accumulation of this set. Thus in order for x to be a point of $u(S_1)$ it is necessary and sufficient that it be a point of $u(S_2)$. This shows that whenever $S \in S_u(x)$ it must be the case that the set $S \cap (x - \delta, x + \delta)$ is also in $S_u(x)$. These details show that S_u is a local system.

Let us now show that S_u has a filtering dual. For this we need to show, because of lemma (5.1), that for any sets S_1 and S_2 with $x \in S_1$, $x \in S_2$, and $S_1 \cup S_2 \in S_u(x)$ it must be the case that one at least of the sets S_1 and S_2 belongs to $S_u(x)$. If not then x does not belong to either of the sets $u(S_1 \setminus \{x\})$ or $u(S_2 \setminus \{x\})$. But
$$u(S_1 \setminus \{x\}) \cup u(S_2 \setminus \{x\}) = u([S_1 \cup S_2] \setminus \{x\})$$
so that x cannot then belong to $u(S_1 \cup S_2 \setminus \{x\})$. This means that the union $S_1 \cup S_2$ cannot belong to $S_u(x)$ in contradiction to the assumption above. This proves then that S_u has the property of lemma (5.1) which is equivalent to the fact of a filtering dual.

Conversely suppose that S is a local system that has a filtering dual. We construct a closure operator v in such a way that $S = S_v$. Define firstly the "derivative" of a set A relative to S by writing
$$(S)\text{-der}(A) = \{x : A \cup \{x\} \in S(x)\}$$
and define v as
$$v(A) = (S)\text{-der}(A) \cup A.$$
We check that v is a closure operator. Only the additive property is not immediate. For this let the sets A_1 and A_2 be given. Certainly

$$v(A_1 \cup A_2) \supset v(A_1) \cup v(A_2).$$
In the other direction if x is a point in $v(A_1 \cup A_2)$ then
$$A_1 \cup A_2 \cup \{x\} \in S(x).$$
Because S has a filtering dual and because of lemma (5.1) we see that one of the sets $A_1 \cup \{x\}$ or $A_2 \cup \{x\}$ must belong to $S(x)$. But this says that x belongs to one of the sets $v(A_1)$ or $v(A_2)$. Thus we must have
$$v(A_1 \cup A_2) \subset v(A_1) \cup v(A_2).$$
This together with the previous set inequality (and an induction argument) shows that v is additive.

Thus the v we have defined is a closure operator. The proof is completed by checking that $S_v(x) = S(x)$ at each point x and this is immediate.

Note that if a closure operator u is the closure operator associated with a topology, i.e. if τ is a topology and $u(A)$, for any set A, is the closure of the set A in the τ-topology, then
$$S_u = (S_\tau)^*.$$
Thus a system based on the closure operator for a topology is the dual of the system based on that topology.

§11. Negligent limits.

The idea of computing the limits of a function in such a way that one ignores some inconvenient set or class of sets goes back at least to Blumberg [22]. The classes of sets that he used in his investigation were the classes of denumerable sets, of first category sets, and of measure zero sets. These limits he termed "approximate" limits but nowadays that term is rather firmly applied to a different process. Cszásár [52] has referred to such limits as "limites sous negligence ... " and, for lack of a better or more standard terminology, we shall call these limits <u>negligent limits</u> where the class of sets that is to be neglected will be assumed to be a given ideal or σ-ideal of sets of real numbers.

The study of derivates taken with negligence of the first category sets was begun by S.Marcus [154], [155], [156]. This was followed by Cszásár's studies [52] and [53] in which derivates are taken with the negligence of an arbitrary ideal [σ-ideal] N of subsets of the real line. That is to say N is a nonempty family of subsets of real numbers with the property that if $\{N_1, N_2, N_3, \ldots\}$ is any finite sequence [resp. infinite sequence] of sets from N then N must contain any subset of their union.

Relative to such an ideal N we may express such limits by the construction of a system S_N and its dual S_N^* as :
$$S_N(x) = \{S : x \in S \text{ and for some } \delta > 0, (x - \delta, x + \delta) \setminus S \in N\},$$
and

$S_N^*(x) = \{ S : x \in S \text{ and for all } \delta > 0, S \cap (x - \delta, x + \delta) \notin N \}$.

These are local systems provided only that the ideal N contains no intervals. The standard choices for N here are the σ-ideals of denumerable sets, of first category sets, of measure zero sets, and of σ-porous sets; but it appears that all of the reasonable theory of such limits requires no particular specialization.

We will simplify the notation by using the expression

$$N - \lim_{y \to x} f(x)$$

for limits taken with respect to this system, and

$$\overline{D}_N f(x), \quad \overline{D}_N^+ f(x), \quad \overline{D}_N^- f(x) \text{ etc.}$$

for the derivates and one-sided derivates taken this way. For convenience of reference we may refer to these as "negligent Dini derivatives". In the special case where N is the σ-ideal of first category sets one commonly calls these the "qualititative Dini derivatives".

§12. Density systems.

Perhaps the most important of the families of local systems is provided by density considerations. In this section we collect some observations on this kind of system. A density system is a local system S such that each $S(x)$ is described as the collection of all sets meeting some density requirement at the point x. Since we may consider either exterior or interior (Lebesgue) measure in the density description, and either upper or lower density there, and choose that density to lie in any range of the interval $[0, 1]$, there are a variety of such systems available.

It would be convenient to have a compendious system of notations that permitted us to refer to each of the possible density systems but it would certainly tax the patience of all but the most generous reader. Instead we shall define anew each density system that will be required, with the exception only of the most important ones. For the interior right density of a set A at a point x (upper and lower) we write

$$\underline{d}_+^i (A, x) = \liminf_{h \to 0+} \frac{|A \cap (x, x + h)|^i}{h}$$

$$\overline{d}_+^i (A, x) = \limsup_{h \to 0+} \frac{|A \cap (x, x + h)|^i}{h}$$

with similar obvious notations for the exterior right densities, the left versions and the bilateral versions.

(12.1) DEFINITION. The <u>approximate system</u> S_{ap} and the <u>weak approximate system</u> S_{wap} are defined as

$$S \in S_{ap}(x) \quad \text{if and only if } x \in S \text{ and } \underline{d}^i (S, x) = 1,$$

and

$$S \in S_{wap}(x) \quad \text{if and only if } x \in S \text{ and } \underline{d}^e (S, x) = 1.$$

Notice that the density system S_{ap} (sets taken with inner, lower density 1) is filtering since, if two sets S_1 and S_2 both have inner lower density 1 at the point x, then so too must their intersection $S_1 \cap S_2$. This would not be the case with any of the other density systems. Indeed this system is precisely that given in example (9.2) as arising from the density topology.

(12.2) LEMMA. The density system S_{ap} is the system given by the density topology τ_d on IR.

PROOF. See Goffman and Waterman [104]. .

The dual of any density system is another density system. Since a set S with $x \in S$ is in the dual $S^*(x)$ of a system S if and only if the complementary set $(IR \setminus S) \cap \{x\}$ is not in the original system $S(x)$ it is easy to see that the dual of any given density system requires an interchange of exterior and interior densities, and of upper and lower densities.

(12.3) THEOREM. In the family of density systems the following dualities must hold.
(i) The dual of S_{ap} is the density system S_1 described by the requirement that $S \in S_1(x)$ if and only if $x \in S$ and either
$$\overline{d}_+e(S, x) > 0 \quad \text{or} \quad \overline{d}_-e(S, x) > 0.$$
(ii) The dual of S_{wap} is the density system S_2 described by the requirement that $S \in S_2(x)$ if and only if $x \in S$ and either
$$\overline{d}_+i(S, x) > 0 \quad \text{or} \quad \overline{d}_-i(S, x) > 0.$$
(iii) For any $t \in [0, 1]$ write S_t and S^t for the density systems desribed by the requirement,
$$S \in S_t \text{ iff } x \in S \text{ and } \underline{d}_+i(S, x) > t \quad (\text{"=" if } t = 1)$$
$$S \in S^t \text{ iff } x \in S \text{ and } \overline{d}_+e(S, x) > t \quad (\text{"=" if } t = 1).$$
Then $(S_t)^* = S^{1-t}$.
(iv) Similarly for any $t \in [0, 1]$ write S_t and S^t for the density systems desribed by the requirement,
$$S \in S_t \text{ iff } x \in S \text{ and } \underline{d}_+e(S, x) > t \quad (\text{"=" if } t = 1)$$
$$S \in S^t \text{ iff } x \in S \text{ and } \overline{d}_+i(S, x) > t \quad (\text{"=" if } t = 1).$$
Then $(S_t)^* = S^{1-t}$.

PROOF. Each of these follows directly from the definition of a dual system.

We shall refer to limits taken with respect to the system S_{ap} as <u>approximate limits</u>,

and we prefer the notation
$$\overline{D}_{ap} f(x) , \underline{D}_{ap} f(x)$$
for the bilateral approximate derivates, and
$$\overline{D}_{ap}{}^+ f(x) , \underline{D}_{ap}{}^+ f(x) , \overline{D}_{ap}{}^- f(x) , \text{ and } \underline{D}_{ap}{}^- f(x)$$
for the four approximate Dini derivatives. In later chapters many results expressible in terms of these derivates will be given; for a recent survey of the whole area of approximate derivatives and their applications see the article of Bruckner and Goffman [37].

The approximate limit is defined commonly in a different manner than we are taking here. In order to clarify the relationship we state a lemma that shows the equivalence. Recall that for a function f at a point x the approximate limit superior, ap-limsup $f(x)$ is usually defined (see Saks [209, p.218]) as the infimum of all numbers t such that the set $\{y : f(y) > t\}$ has x as a point of dispersion (i.e. so that this set has outer upper density 0 at the point x). Similarly the approximate limit inferior is defined. Then the approximate limit is taken as the common value of these two extreme limits should they be the same.

(12.4) LEMMA. For an arbitrary function f at a point x the extreme approximate limits and the approximate limits may be written as
$$\text{ap-limsup } f(x) = (S_{ap})\text{-limsup } f(x) = (S_1)\text{-liminf } f(x) ,$$
$$\text{ap-liminf } f(x) = (S_{ap})\text{-liminf } f(x) = (S_1)\text{-limsup } f(x) ,$$
and,
$$\text{ap-lim } f(x) = (S_{ap})\text{-lim } f(x) ,$$
where $S_1 = S_{ap}{}^*$ is as in theorem (12.3), (i).

§13. Miscellaneous limits.

There are a number of special examples that are of use from time to time. In this section we present a few of these, partly for historical interest and partly to illustrate further the diversity of ideas.

(13.1) Example. [Positive measure system] We may define a positive measure system S by requiring that for each x the family $S(x)$ contains any set that has always positive measure in every neighbourhood of x. Since we may use either exterior or interior measure this gives us two systems:
$$S_e(x) = \{ S : x \in S \text{ and } |S \cap (x-\delta, x+\delta)|^e > 0 \text{ for all } \delta > 0 \} ,$$
and
$$S_i(x) = \{ S : x \in S \text{ and } |S \cap (x-\delta, x+\delta)|^i > 0 \text{ for all } \delta > 0 \} .$$
Similarly we may define one sided versions of these two systems in the obvious manner, or require positive measure on both sides of the point x. See Theorem (29.1) for an application of this system in the setting of cluster sets.

(13.2) Example. [Full measure system] In a similar fashion we define a closely related system by using full measure in an interval. We let S denote the system defined so that $S \in S(x)$ if and only if $x \in S$ and there is some $\delta > 0$ so that the set $S \cap (x - \delta, x + \delta)$ has full exterior measure in the interval $(x - \delta, x + \delta)$.

Note that there is a dual relation between systems of this type and systems decribed in the preceding example.

(13.3) Example. [Congruent path system] Suppose that a fixed set Q is given that contains 0 and has 0 also as a point of accumulation. At each point x we define the system S_Q by writing

$$S_Q(x) = [\, S : S \supset (Q + x) \cap (x - \delta, x + \delta) \text{ for some } \delta > 0 \,].$$

Then S_Q is a local system that we shall call the congruent path system based on the set Q. Note that the assertion

$$(S_Q)\text{-lim } f(x) = c$$

means exactly that

$$\lim_{t \in Q, t \to 0} f(x + t) = c$$

so that S_Q limits are just path limits along translates of the basic set Q. These ideas have been extensively studied by Sindalovskiĭ [214], [215], [216], [217], [218], [219], [220], [221] and Starcev [223].

(13.4) Example. [Congruent sequential path system] Let there be given a sequence of points $\{q_n\}$ converging to zero, and define the set

$$Q = \{0\} \cup \{\, q_n : n = 1, 2, 3, \ldots \,\}.$$

Then the congruent system defined using this set Q gives the sequential limit studied by Laczkovich and Petruska [145], [146]; that is

$$(C_Q)\text{-lim } f(x) = c$$

if and only if

$$\lim_{n \to \infty} f(x + q_n) = c.$$

(13.5) Example. [Path systems] Suppose that at every point of the real line there has been given a set E_x with the property that $x \in E_x$ and x is also a point of accumulation of E_x. Such a set we shall call a <u>path leading to</u> $\underline{x}$. We define relative to the system of paths

$$E = \{\, E_x : x \in \mathbb{R} \,\}$$

a local system S_E so that at each point $x \in \mathbb{R}$, $S_E(x)$ is the filter generated by the filterbase

$$\{\, E_x \cap (x - \delta, x + \delta) : \delta > 0 \,\}.$$

This system S_E defines limits along the paths in E. In order for $(S_E) \lim f(x) = c$ one

has
$$\lim_{y \in E_x, y \to x} f(y) = c.$$

Such limits are studied in Bruckner, O'Malley and Thomson [43] as path limits; this study presented a preliminary version of the considerations developed in this monograph in greater detail and in greater generality. Note that path limits include the congruent and sequential limits as special cases, but that general limits using local systems are more general still. In section 14 we will discuss the exact connection between path type limits and system limits.

(13.6) Example. [Selective systems] By a *selection* p is meant a function $p(x,y)$ of two real variables with the properties:

(i) $p(x,y) = p(y,x)$, and

(ii) if $x < y$ then $x < p(x,y) < y$.

Following O'Malley [180] we define a selective limit relative to a given selection p as
$$(p)\text{-}\lim f(x) = \lim_{y \to x} f(p(x,y)).$$

This limit can be expressed in the language of local system limits by constructing, relative to a selection p, the system S_p defined by setting $S_p(x)$ as the collection of all sets S such that $x \in S$ and for some $\delta > 0$,
$$S \supset \{ p(x,y) : 0 < |y-x| < \delta \}.$$

(13.7) Example. [Porosity systems] There are a number of problems in differentiation theory that might be described in terms of some system that is based on the notion of porosity. Here we present a family of local systems that utilize the various porosity computations.

For any number $0 \leq t < 1$ we write the porosity system S^t by defining
$$S^t(x) = \{ S : x \in S, \text{ and the porosity of } S \text{ at } x \text{ is } \leq t \}.$$
To include the extreme case $t = 1$ we just include sets that are not bilaterally strongly porous:
$$S^1(x) = \{ S : x \in S, S \text{ has left or right porosity at } x < 1 \}.$$
Similar definitions permit one-sided versions of the porosity systems.

Occasionally one might need some refinements on the porosity zero systems. For these we can sharpen the porosity computation by using the notion of ψ-porosity as described in the appendix. A number of results in differentiation theory have been expressed in terms of "porosity limits" and "porosity derivatives" in Bruckner and Thomson [44].

(13.8) Example. [Quasi limits] Following Kempisty [134] let us say that a function f has a *quasi limit* c at a point x if for every neighbourhood U of c and every neighbourhood V of x there is an open set V_1 contained in V such that $f(V_1) \subset U$.

Such a limit may be expressed as a limit taken relative to an appropriate local system. Indeed at each x we define $S(x)$ to be the family of sets S such that $x \in S$ and S has a nonenmpty interior. Such systems play a role in (35.6) for example. For a study of quasi limits or of functions continuous in such a sense see Bledsoe [20], Ewert and Lipiński [76], Marcus [157], Neubrunnova [170], and Theilman [231].

§14. Path approach.

There is a distinction in our setting between a system limit and a limit along a set. Suppose, for example, that a system S has been given and that a function f has the property that at a point x_0 the limit

$$\lim_{y \to x_0, y \in S} f(y) = c$$

exists and gives this number c for at least one choice of a set $S \in S(x_0)$. Then it must certainly be the case that the system limit relative to the system S at the point x_0 is the same, i.e. that

$$(S)\text{-lim } f(x) = c.$$

We need observe only that the set $\{ t : |f(t) - c| < \epsilon \}$ must be, for every $\epsilon > 0$, in the family $S(x_0)$ since it contains at least a portion of the set S near to x.

Consequently when "path limits" along any set in the system S exist the general system limit must itself exist. The converse need not however happen. A system limit may exist and yet for no set S in the system does the limit along the path S exist. The examples show how this may happen and give the clue as to how this situation may be characterized.

(14.1) Example. The density system S_{ap} (sets having interior, lower density 1) has this path approach property, whereas for other density families it may fail. Suppose that it is the case that a function f has a limit in this sense, say that

$$(ap)\text{-lim } f(x) = c.$$

Then it is possible to find a set $S \in S_{ap}(x)$ so that in fact

$$\lim_{y \in S, y \to x} f(y) = c.$$

Thus, in this case, a limit relative to the system is in fact a limit along one of the sets in the system.

To see how this set S may be constructed we define firstly the sets

$$A_n = \{ y : |f(y) - c| < \frac{1}{n} \}$$

which, by definition, must be members of $S_{ap}(x)$, so that each set A_n has interior density 1 at x. From these sets A_n we fashion a path leading to x that has interior density 1 at x and along which the limit of f is c. We give the construction just to produce the set on the right of the point x as this is sufficient to illustrate the methods. Choose a decreasing sequence $\{h_n\}$ tending to zero in such a way that

$$|A_n \cap (x, x+t)|^i > (1 - \frac{1}{2n})t$$

for $0 < t < h_n$ and choose a further sequence $\{\tau_n\}$ tending to zero and sufficiently small that

$$|A_n \cap (x+\tau_n, x+t)|^i > (1 - \frac{1}{n})t$$

for $h_{n+1} \leq t < h_n$.

Define the set

$$S = \sum_{1}^{\infty} A_n \cap (x+\tau_n, +\infty).$$

One then checks easily that S has right interior density 1 at x and that the limit of f along this set S at x is the required number c.

(14.2) Example. Let now S be the density system described by the requirement that each set $S \in S(x)$ have right lower (exterior) density greater than zero at x. In contrast to the previous example we construct an example of a function f for which it is true that

$$(S)\text{-lim } f(0) = 0$$

exists but, nonetheless, there is no set S having right lower (exterior) density positive so that

$$\lim_{y \in S, y \to 0} f(y) = 0.$$

Take $\{S_n\}$ a decreasing sequence of sets having density $1/n$ at 0 and define the function f to have the value $1/n$ for $x \in S_n \setminus S_{n+1}$ and to have the value 2 at every other point. One checks that the system limit is zero but that there is no path having lower density positive along which this limit is zero.

The hypothesis that is required on a system in order that it have the property that all limit numbers using sets in the system are actually path approached along sets in the system is just a formalization of the computations that appear in example (14.1) above for the density 1 systems. This was first given an abstract expression by Jędrzejewski [129]. The condition (b) of this theorem will be referred to as $[J_2]$ in the sequel. The equivalent condition (c) was used by Tevy and Bruteanu [230]. Similar considerations appear in a different, but related, abstract setting in Agronsky [2] and Nishiura [173].

(14.3) THEOREM. Let S be a local system. Then at a point x the following three properties are equivalent:

(a) For any point function f if $(S)\text{-lim } f(x) = c$ then there is a set S belonging to $S(x)$ so that
$$\lim_{y \in S, y \to x} f(y) = c.$$

(b) The family $S(x)$ has the following property $([J_2])$: for any decreasing sequence of sets $\{S_n\}$ belonging to $S(x)$, there must be a sequence of positive numbers $\{\tau_n\}$ decreasing to zero so that the set
$$\{x\} \cup \left[\sum_{n=1}^{\infty} [S_n \setminus (x - \tau_n, x + \tau_n)] \right]$$
belongs to $S(x)$.

(c) The family $S(x)$ has the property that, for any decreasing sequence of sets $\{S_n\}$ belonging to $S(x)$, there must be a sequence of positive numbers $\{\tau_n\}$ decreasing to zero and a set $S \in S(x)$ so that for each index n,
$$S \cap (x - \tau_n, x + \tau_n) \subset S_n.$$

PROOF. Suppose that the property (b) of the theorem does hold and that
$$(S)\text{-lim } f(x) = c$$
for some function f. We shall show how (b) may be used to construct a set S in $S(x)$ so that
$$\lim_{y \in S, y \to x} f(y) = c.$$
For convenience we take it that c is finite although a similar construction will work if $c = +\infty$ or $c = -\infty$. Define the sequence of sets
$$S_n = \{y : y = x \text{ or } |f(y) - c| < \tfrac{1}{n}\},$$
each of which is evidently in $S(x)$. By the hypothesis (b) there must be a sequence of numbers $\{\tau_n\}$ decreasing to zero and with the property that the set
$$S = \{x\} \cup \left[\sum_{n=1}^{\infty} [S(n) \setminus (x - \tau_n, x + \tau_n)] \right]$$
belongs to $S(x)$. By the way in which this set has been constructed it should be evident that
$$\lim_{y \in S, y \to x} f(y) = c,$$
exactly as required to establish that (b) implies (a).

In the other direction suppose that the condition (b) of the theorem fails; then we shall exhibit a function f and a number c for which there is no path limit within the system. Let us suppose that there is given a decreasing sequence $\{S_n\}$ of sets from $S(x)$ such that for no sequence $\{\tau_n\}$ descending to zero does the condition of the theorem

hold. For convenience we take it that $x = 0$ and we will design a function f for which
$$(S)\text{-lim}\, f(0) = 0 \ .$$
Define $f(t)$ to be $1/n$ at points t in $S_n \setminus S_{n+1}$ and to be 2 at any remaining points. It is clear that the S-limit of f at $x = 0$ is zero and we show that $S(0)$ can contain no set S along which the limit of f at 0 is zero.

Suppose, contrary to this, that there is such a set S. If so then there must be a sequence of numbers $\{\tau_n\}$ descending to zero so that, whenever
$$t \in S \cap (-\tau_n, \tau_n),$$
one has
$$|f(t) - 0| < \frac{1}{n} \ .$$
This then gives that the set $S \cap (-\tau_1, \tau_1)$ is
$$[\,\{0\} \cup S\,] \cap \left(\sum_{n=1}^{\infty} (-\tau_n, \tau_n) \setminus (-\tau_{n+1}, \tau_{n+1}) \right)$$
which is a subset of the set
$$\{0\} \cup \left(\sum_{n=1}^{\infty} [S_n \setminus (-\tau_n, \tau_n)] \right).$$
But this then forces the set
$$\{0\} \cup \left(\sum_{n=1}^{\infty} [S_n \setminus (-\tau_n, \tau_n)] \right)$$
to belong to $S(0)$ contradicting our earlier assumption. From this contradiction we obtain that (a) implies (b). Finally the equivalence of condition (c) of the theorem with (a) and (b) may be similarly obtained.

This property of Jędrzejewski may also be used to characterize those systems that are identical to the path systems of example (13.5). Let us say that a system S is **path generated** at a point x if the family $S(x)$ is the filter generated by a filterbase of the form
$$\{\, E \cap (x - \epsilon, x + \epsilon) : \epsilon > 0 \,\}$$
for some set E. Note that this will require that E contain x and have x as a point of accumulation.

(14.4) THEOREM. A necessary and sufficient condition that a system $S(x)$ at a point x be path generated is that $S(x)$ be a filter that has a countable base and that satisfies any of the three equivalent hypotheses of theorem (14.3).

PROOF. It is clear that the conditions are necessary and so we need to prove that they are

sufficient. Suppose that $\{S_1, S_2, S_3, \ldots\}$ is a sequence that forms a countable base for the filter $S(x)$. We write
$$T_n = S_1 \cap S_2 \cap S_3 \ldots \cap S_n$$
and apply the condition (b) of theorem (14.3) to obtain a sequence $\{\tau_n\}$ and a set
$$S = \{x\} \cup \left(\sum_{n=1}^{\infty} [T_n \setminus (x - \tau_n, x + \tau_n)] \right)$$
that belongs to $S(x)$. For this set S and for any set $S_x \in S(x)$ there must be, by the construction and by the fact that the sets $\{S_n\}$ form a base for the filter $S(x)$, a set T_n with $S_x \supset T_n$. Consequently
$$S_x \supset S \cap (x - \tau_n, x + \tau_n).$$
This then shows that $S(x)$ is exactly the filter generated by the path S, i.e. $S(x)$ is generated by the filterbase
$$\{S \cap (x - h, x + h) : h > 0\}$$
and the theorem is proved.

(14.5) Example. The density systems S_{ap} (inner, density 1) and S_{wap} (exterior density 1) have the property expressed in theorem (14.3). The other density systems do not share this property. The computations were given above. Note that neither is, however, path generated.

(14.6) Example. If N is a σ-ideal of subsets of the real line and S_N is the system generated as in section 11 then one can see that it must satisfy the path property given in the theorem. Again note that such a system may not be countably generated so that it need not itself be path generated although limits taken relative to such a system are path limits within the system.

§15. Intersection conditions.

The local assumptions introduced in the preceding sections play a key role in many investigations. Certain results, however, can not be proved on the basis of an assumption about the nature of the sets in each family $S(x)$, but require some information about the way sets in neighbouring families $S(x)$ and $S(y)$ interact. To obtain such an interconnection between points x and y that are considered "close" we need to have some information about the nature of the intersection $S_x \cap S_y$ for pairs of sets $S_x \in S(x)$ and $S_y \in S(y)$. By an intersection condition on a set X for a system S we mean an hypothesis that for every choice of sets

$$\{S_x : x \in X\},$$
with each $S_x \in S(x)$, there is a positive function δ on X so that whenever
$$0 < y - x < \min\{\delta(x), \delta(y)\}$$
it is the case that the sets S_x and S_y intersect in some specified fashion.

We shall require a variety of intersection conditions in the sequel; some such are listed below for illustration. The idea appears formally for the first time in Bruckner, O'Malley and Thomson [43]. Other types of intersection conditions have been used in O'Malley [188] and we shall use a variety of these conditions in this study.

(15.1) $\qquad S_x \cap S_y \neq \emptyset$.

(15.2) $\qquad S_x \cap S_y \cap [x, y] \neq \emptyset$.

(15.3) $\qquad S_x \cap S_y \cap (x, y) \neq \emptyset$.

(15.4) $\qquad S_x \cap S_y$ contains a point other than x or y.

(15.5) $\qquad S_x \cap S_y \cap [x - \lambda(y-x), x] \neq \emptyset$
and/or
$\qquad S_x \cap S_y \cap [y, y + \lambda(y-x)] \neq \emptyset$

where the parameter λ ($0 < \lambda < +\infty$) is specified.

The intersection conditions are used mainly to induce a partition of the set X on which they hold. Since this argument will be used with great frequency in the ensuing material we pause to formalize it. This is just a general account of a type of argument that has long been used in analysis but without explicit comment; the first attempt to describe systematically the details in some general language appears in Bruckner, O'Malley and Thomson [43] under the terminology "δ-decomposition".

Suppose that a positive function δ has been defined on a set X. We shall refer to δ as a <u>guage</u> on X (as do a number of authors in a different setting). Then we say that the guage δ <u>induces a partition</u> of X into a sequence of sets $\{X_{nm}\}$ ($n = 1, 2, 3, \ldots$, $m = 0, \pm 1, \pm 2, \pm 3, \ldots$) as follows: we let firstly $\{X_n\}$ denote the collection
$$X_n = \left\{ x \in X : \delta(x) \in \left(\frac{1}{n}, \frac{1}{n-1}\right] \right\}$$
and let $\{X_{nm}\}$ denote the collection
$$X_{nm} = \left\{ x \in X_n : x \in \left(\frac{m-1}{n}, \frac{m}{n}\right] \right\}$$
for $n = 1, 2, 3, \ldots$ and $m = 0, \pm 1, \pm 2, \pm 3, \ldots$.

The $\{X_{nm}\}$ is a denumerable partition of X that has the following features:

(i) if x and y belong to the same member of the partition then necessarily
$$|x-y| < \min\{\delta(x), \delta(y)\} .$$
(ii) if $x \in X$ is in the closure of a set X_{nm} and y is in X_{nm} with y required to lie in the interval $(x - \delta(x), x + \delta(x))$ then again one must have
$$|x-y| < \min\{\delta(x), \delta(y)\} .$$
(iii) if δ is a Borel [measurable] function and X a Borel [measurable] set then the members of the partition are Borel [measurable] sets.

The fact that such a function δ induces a denumerable partition of the set X then allows a great many arguments to be used. For example, to show that a set X is small in some sense, say in the sense of measure, one can use the function δ to induce a partition of the set X and then argue that each member of the partition has measure zero. The feature (ii) that permits one to draw in points that are in the closure of a member of the partition is most useful in applications that use Baire's theorem. These ideas are, of course, somewhat technical but, after some very brief familiarity with the general pattern, one can invoke this argument quite routinely.

The intersection conditions are used throughout the work only to induce a partition of some set. Indeed the existence of such a partition is equivalent to the intersection condition and the definitions could well have been framed about the existence of a partition, rather than the existence of the gauge function δ. To see this, suppose such a partition is available. Then we may construct an appropriate function δ on X so that the intersection condition is available. For convenience suppose the partition is written as a singly indexed sequence $\{Y_n\}$ and define $\delta(x) = 1$ for $x \in Y_1$; for x in $\overline{Y}_1$ but not in Y_1 (with the closure taken just inside the set X) define $\delta(x)$ so that any points y from Y_1 that are inside the interval $(x - \delta(x), x + \delta(x))$ permit an intersection for the sets S_x and S_y. Inductively then we define $\delta(x)$ for $x \in Y_n$ and $x \notin \overline{Y}_{n-1}$ so that $\delta(x)$ is less than the distance from x to $\overline{Y}_{n-1}$; and for x in $\overline{Y}_n$ but not in Y_n we take $\delta(x)$ sufficiently small so that any points y that are in both Y_n and $(x - \delta(x), x + \delta(x))$ permit an intersection for the sets S_x and S_y. By such a process the function δ is constructed on the basis of the partition and δ will permit an intersection condition as described above to hold.

We conclude this section by proving that a number of our examples possess certain of these intersection conditions. For application purposes the most important of the intersection conditions that we can discover are those enjoyed by the density systems.

(15.6) LEMMA. Let S^t denote the density system defined by requiring that $S^t(x)$ contain every set that contains x and has lower interior density on each side exceeding t. Then, provided $t > 1/2$ and $\frac{\lambda}{1 + 2\lambda} > 1 - t$, the system satisfies an intersection condition for which $S_x \cap S_y \cap (x, y)$ has positive exterior measure, and also an intersection condition for which both sets
$$S_x \cap S_y \cap (y, y + \lambda(y-x)) \quad \text{and} \quad S_x \cap S_y \cap (x - \lambda(y-x), x)$$
have positive exterior measure.

PROOF. For any choice $\{ S_x : x \in X \}$ from the density system S^t ($t > 1/2$) we may select a positive function δ on X so that
$$\left| S_x \cap (x, y) \right|^i > t |y - x| > \frac{1}{2} |y - x|,$$
for $-\delta(x) < y - x < \delta(x)$. If
$$0 < y - x < \min \{ \delta(x), \delta(y) \}$$
then
$$y - x \leq \left| (x, y) \setminus S_x \right|^e + \left| (x, y) \setminus S_y \right|^e + \left| (x, y) \cap S_x \cap S_y \right|^e.$$
Using the estimates on the measures of these sets we obtain
$$\left| (x, y) \cap S_x \cap S_y \right|^e \geq (1 - 2t)(y - x) > 0$$
and this is the required intersection condition.

For the other conditions (let us consider only the right hand version) suppose that a choice $\{ S_x : x \in X \}$ is given and choose a guage δ on X so that
$$\left| S_x \cap (x, y) \right|^i > t(y - x)$$
whenever $0 < y - x < (\lambda + 1)\delta(x)$. Now, if
$$0 < y - x < \min \{ \delta(x), \delta(y) \},$$
then
$$\lambda(y - x) \leq \left| (y, y + \lambda(y - x)) \setminus S_x \right|^e + \left| (y, y + \lambda(y - x)) \setminus S_y \right|^e$$
$$+ \left| (y, y + \lambda(y - x)) \cap S_x \cap S_y \right|^e.$$
Now, using the estimates
$$\left| (y, y + \lambda(y - x)) \setminus S_x \right|^e \leq (1 - t)(1 + \lambda)(y - x)$$
and
$$\left| (y, y + \lambda(y - x)) \setminus S_y \right|^e \leq (1 - t)\lambda(y - x),$$
we obtain
$$\lambda(y - x) \leq (1 - t)(1 + \lambda)(y - x) + (1 - t)\lambda(y - x) + \left| (y, y + \lambda(y - x)) \cap S_x \cap S_y \right|^e.$$
Thus
$$\left| (y, y + \lambda(y - x)) \cap S_x \cap S_y \right|^e \geq \left[\frac{\lambda}{1 + 2\lambda} - (1 - t) \right](1 + 2\lambda)(y - x)$$
which gives the required intersection condition. Note that the arithmetic requires $t > 1/2$.

(15.7) LEMMA. Let N be an ideal of subsets of the real line and let S_N denote the associated system. Then S satisfies an intersection condition of the form
$$S_x \cap S_y \supset [x - (y - x), y + (y + x)] \setminus N$$
for some set N in the ideal N.

PROOF. Given a choice $\{S_x : x \in X\}$ one selects a positive function δ on the set X so that
$$[x - 3\delta(x), x + 3\delta(x)] \setminus S_x = N_x \in S_N(x).$$
Then given any points x and y from X for which
$$0 < h = y - x < \min\{\delta(x), \delta(y)\}$$
it is clear that
$$S_x \cap S_y \supset [x - h, y + h] \setminus (N_x \cup N_y)$$
which gives the required intersection condition.

(15.8) Example. For the systems S_0, S_0^+, and S_0^- the following intersection conditions may be simply proved. For S_0 an intersection condition of the form
$$S_x \cap S_y \supset [x - (y - x), y + (y - x)]$$
may be proved. Similarly for S_0^+ one has an intersection condition of the form
$$S_x \cap S_y \supset [y, y + (y - x)].$$

§16. S-covers.

The notion of a covering of a set by a family of intervals plays an important role in many investigations in differentiation theory. To see how this concept has arisen in such studies we look at an elementary example. Suppose that at each point x of a set X one has the inequalities
$$c_1 < \underline{D} f(x) < c_2$$
for the lower bilateral derivate of the function f. Then one can collect all the intervals $I = [x, y]$ or $[y, x]$ for which
$$\frac{f(y) - f(x)}{y - x} > c_1$$
into a family C_1 and similarly one can form the family C_2 of all the intervals $I = [x, y]$ or $[y, x]$ for which
$$\frac{f(y) - f(x)}{y - x} < c_2.$$

The family C_1 has the property that at every point $x \in X$ there is an abundance of intervals $[x, y]$; in fact there must be a positive number δ so that for every $0 < |y - x| < \delta$ each interval $[x, y]$ (or $[y, x]$) belongs to C_1. Because of this property the family C_1 is said to be a <u>full cover</u> of the set X. Similarly the family C_2 has a kind of dual property: at each point $x \in X$ and for every $\epsilon > 0$ there must be an interval $[x, y]$

or $[y, x]$ in C_2 for which $0 < |y - x| < \epsilon$. Such a family C_2 is said to be a <u>Vitali cover</u> of the set X.

These notions generalize easily to the setting of a local system and they will play an important role in the study of the differentiation theory that arises in such a setting.

(16.1) DEFINITION. Let S be a local system on a set X and let C be a family of intervals. We say that C is an S-<u>cover</u> of the set X if for every point $x \in X$ the set
$$\{ y : y = x \text{ or } y > x \text{ and } [x, y] \in C, \text{ or } y < x \text{ and } [y, x] \in C \}$$
belongs to the family $S(x)$.

The examples S_0 and S_∞ at the two extremes of the lattice of local systems indicate the scope of the idea.

(16.2) Example. Consider the system S_0 and its dual S_∞. In order for C to be a S_0-cover of a set X there must be a guage δ on X so that for every $0 < y - x < \delta(x)$ the interval $[x, y] \in C$ and for every $0 < x - y < \delta(x)$ the interval $[y, x] \in C$. Such a cover in the literature is called a <u>full cover</u> of the set X, and consequently the notion of an S_0-cover is equivalent to that.

At the other extreme in order for C to be an S_∞-cover of X it must be the case that for every $x \in X$ and any positive number δ there is a number y in the interval $(x - \delta, x)$ or in the interval $(x, x + \delta)$ so that the interval $[y, x]$ or $[x, y]$ belongs to C. Such a cover in the literature is called a <u>Vitali cover</u> of the set X, and consequently the notion of an S_∞-cover is essentially the classical notion of a Vitali cover.

In particular we see that any S-cover lies between these two extremes; thus any S-cover of a set X is also a Vitali cover of that set.

(16.3) LEMMA. For any system S if C is an S-cover of a set X then C is a Vitali cover of X.

Our main results for S-covers allow us to extract finite collections of nonoverlapping intervals that have certain properties.

(16.4) LEMMA. Let S be a local system that has an intersection condition of the form
$$S_x \cap S_y \cap [x, y] \neq \emptyset$$
on a set X, and let C be an S-cover of the set X. Then there is a denumerable partition $\{ X_n \}$ of the set X such that, given any pair of points x and y ($x < y$) in a set X_n, C contains a partition of the interval $[x, y]$.

PROOF. Let $\{ S_x : x \in X \}$ be the choice from S that is associated with the cover C and let $\{X_n\}$ be the partition of X that is induced by this choice in accordance with the intersection condition. Then for any points x and y from the same set X_n there must be a point z in the interval $[x, y]$ that belongs to both of the sets S_x and S_y. If z is x or y then the interval $[x, y]$ itself belongs to C and this supplies the partition. Otherwise both of the intervals

$$I_1 = [x, z] \text{ and } I_2 = [z, y]$$

belong to C and again the partition is supplied.

Note that this same construction applies to points x or y that are in the closure of some set X_n and also in X. We assert this in a lemma which follows from the same partitioning argument.

(16.5) LEMMA. Let S be a local system that has an intersection condition of the form

$$S_x \cap S_y \cap [x, y] \neq \emptyset$$

on a closed set X, and let C be an S-cover of the set X. Then there is a denumerable partition $\{X_n\}$ of the set X such that, given any point x in the closure of a set X_n, C contains a partition of the interval $[x, y]$ for all y in X_n that are sufficiently close to x.

From these lemmas it is easy to produce the following theorem that asserts that in many instances an S-cover of an interval must contain a partition of that interval. Here and elsewhere a system S is <u>bilateral</u> provided every set S in $S(x)$ contains points on either side of x.

(16.6) THEOREM. Let S be a local system that is bilateral and has an intersection condition of the form

$$S_x \cap S_y \cap [x, y] \neq \emptyset .$$

Then, if C is an S-cover of the real line, there is in C a partition of any interval $[a, b]$, i.e. for each interval $[a, b]$ there must be a collection

$$\{I_1, I_2, I_3, \ldots, I_k\} \subset C$$

where for $i \neq j$ the intervals I_i and I_j do not overlap and where

$$[a, b] = \sum_{i=1}^{k} I_i .$$

PROOF. The proof follows from the two lemmas by a standard category argument which we sketch here. Let I be the collection of all intervals $[a, b]$ for which a partition does exist for every subinterval of $[a, b]$. We may write

$$G = \sum \{ (a,b) : [a,b] \in I \}.$$

This set G is open and, by the way it has been defined and by a simple compactness argument, we can see that there must be a partition of any interval [c,d] contained entirely in G. Indeed because of the fact that S is bilateral there must be a partition of any interval [c,d] for which (c,d) is a component of G.

Define Q as the complement of G. If Q is empty the proof is completed. Otherwise Q must be a perfect set. The lemma (16.4) above allows us to find a denumerable partition $\{Q_n\}$ of the set Q with that property asserted in the lemma. Using Baire's theorem we may take a set Q_m that is dense in some portion of Q, say Q_m is dense in Q ∩ (c,d).

We leave it to the reader to apply the two lemmas (16.4) and (16.5) to verify that a partition of each subinterval of (c,d) may be found from the cover C. But this is a contradiction. Thus the set Q must have been empty and the theorem is proved.

§17. Variation of a function.

The notion of the variation of a function is due to Camille Jordan in 1881. It arose in his study of rectifiable arcs and in his study of Fourier series, but the idea has very broad applications. It was Lebesgue who showed that functions of bounded variation play important roles in certain problems in integration and differentiation theory. Since then the idea has been carried much further, in one dimension notably by Denjoy and Lusin.

A convenient way of exploring the variation of a function and also of incorporating the more general variational ideas of Denjoy and Lusin was introduced by Henstock [112]. We associate a measure μ_f with a function f in such a way that for a set $X \subset \mathbb{R}$ the measure $\mu_f(X)$ carries some information as to the variation of the function f on the set X. This represents a more subtle and useful tool than the standard Lebesgue-Stieltjes measure since that measure applies only to functions of bounded variation and this notion may be meaningfully applied to much more general functions. In this section we present an introduction to the basic ideas; the full development is given in chapter four.

The definitions needed are listed here.

(17.1) DEFINITION. Let ψ be an interval function and let C be a nonempty family of intervals. By the <u>variation</u> of ψ on C we denote

$$\text{Var}(\psi, C) = \sup \sum_{i=1}^{n} |\psi(I_i)|$$

where the supremum is taken with regard to all sequences of intervals $\{ I_1, I_2, I_3 \ldots \}$

contained in C and for which if $i \neq j$ the intervals I_i and I_j do not overlap. For convenience if C is empty we define
$$\text{Var}(\psi, \emptyset) = 0 .$$

(17.2) DEFINITION. Let S be a local system of sets and let ψ be an interval function. Then the S-variation of ψ on a set X is taken as
$$V(\psi, S; X) = \inf \{ \text{Var}(\psi, C) : C \text{ an } S\text{-cover of } X \} .$$

From this definition there arises a number of different ideas and notations.

(17.3)   If ψ is an interval function and S a local system then the set function $(S)\text{-}\mu_\psi$ defined by
$$(S)\text{-}\mu_\psi(X) = V(\psi, S; X) \qquad (X \subset \mathbb{R})$$
is called the <u>Stieltjes measure</u> associated with ψ and S.

(17.4)   If f is a real function and Δf denotes the usual interval function
$$\Delta f([x, y]) = f(y) - f(x) \qquad (x < y)$$
then the set function $(S)\text{-}\mu_f$ defined by
$$(S)\text{-}\mu_f(X) = V(\Delta f, S; X)$$
is called the <u>Stieltjes</u> measure associated with the function f for the system S.

(17.5)   A function f is said to have <u>zero</u> S-variation on a set X if the measure $(S)\text{-}\mu_f$ vanishes on X.

(17.6)   A function f is said to have <u>finite</u> S-variation on a set X if the measure $(S)\text{-}\mu_f$ is finite on X.

(17.7)   A function f is said to have σ-<u>finite</u> S-variation on a set X if the measure $(S)\text{-}\mu_f$ is σ-finite on X.

(17.8)   Two interval functions ψ_1 and ψ_2 are said to be S-<u>variationally equivalent</u> on a set X provided
$$V(\psi_1 - \psi_2, S; X) = 0 .$$

§18. The variational measure.

Our basic theorem asserts that the measures introduced above in (17.3) are genuine outer measures on the real line. Moreover these measures are *metric* outer measures which means that each Borel set is measurable with respect to any one of these measures. These facts will be quite useful in any applications of these ideas. For an elementary account of the notions of outer measure and metric outer measure and the development of these ideas the reader is referred to Munroe [167] or Rogers [206].

(18.1) THEOREM. Let S be an arbitrary local system and ψ an interval function. Then $(S)\text{-}\mu_\psi$ is a metric outer measure on the real line.

PROOF. Let $X, Y_1, Y_2, Y_3, \ldots$ be a sequence of sets for which
$$X \subset \sum_{i=1}^{\infty} Y_i.$$
Suppose that a positive number ε has been given and choose S-covers C_i of the sets Y_i in such a way that
$$\text{Var}(\psi, C_i) \leq \mu_\psi(Y_i) + \frac{\varepsilon}{2^i}.$$
If we define C as the union of the families C_i then it is clear that C is an S-cover of the set X and that therefore
$$\mu_\psi(X) \leq \text{Var}(\psi, C) \leq \sum_{i=1}^{\infty} \text{Var}(\psi, C_i)$$
$$\leq \left[\sum_{i=1}^{\infty} \mu_\psi(Y_i) \right] + \varepsilon.$$
As ε is an arbitrary positive number we have then the inequality
$$\mu_\psi(X) \leq \sum_{i=1}^{\infty} \mu_\psi(Y_i)$$
as required to establish that μ_ψ is an outer measure.

To see that μ_ψ is a metric outer measure suppose that two sets X_1 and X_2 are separated in such a way that there are open sets G_1 and G_2 with $X_1 \subset G_1$, $X_2 \subset G_2$, and $G_1 \cap G_2 = \emptyset$. Then if C is an S-cover of the set $X_1 \cup X_2$ chosen in such a way that
$$\text{Var}(\psi, C) \leq \mu_\psi(X_1 \cup X_2) + \varepsilon$$
we may define the collections C_1 and C_2 by writing
$$C_i = \{ I \in C : I \subset G_i \} \qquad (i = 1, 2).$$
Now we compute (with obvious justifications)

$$\mu_\psi(X_1) + \mu_\psi(X_2) \leq \text{Var}(\psi, C_1) + \text{Var}(\psi, C_2)$$
$$\leq \text{Var}(\psi, C) \leq \mu_\psi(X_1 \cup X_2) + \epsilon.$$

Since $\epsilon > 0$ is arbitrary we must have the equality
$$\mu_\psi(X_1 \cup X_2) = \mu_\psi(X_1) + \mu_\psi(X_2)$$
which is precisely what is required in order to show that μ_ψ is a metric outer measure.

§19. The increasing sets property.

In this section we prove that the variational measures have a property that is known in the literature as "the increasing sets property". If X_n is an expanding sequence of sets and μ an outer measure then μ is said to have the increasing sets property if

$$\lim_{n \to \infty} \mu(X_n) = \mu\left(\sum_{n=1}^{\infty} X_n\right).$$

Our result asserts that, under mild hypotheses on the local system S, the outer measures (S)-μ_ψ will have this property.

(19.1) THEOREM. Let S be a local system that is filtering. Then for any interval function ψ the outer measure (S)-μ_ψ has the increasing sets property.

PROOF. This follows from a series of lemmas which we now prove. We need a type of "uniform" property of the variation relative to intervals and to express this let us introduce some notation. For any set K of real numbers we write the variation relative to K by a slight modification of the definition (17.1) above:
$$\text{Var}_K(\psi, C) = \sup \{ \Sigma \, |\psi(I_i)| : I_i \subset K, I_i \in C \}$$
where $\{I_i\}$ is a sequence of nonoverlapping subintervals of K. Then we define
$$V_K(\psi, S; X) = \inf \{ \text{Var}_K(\psi, C) : C \text{ an } S\text{-cover of } X \}$$
so that this variation relative to K is merely computed using only intervals that are subsets of K. Our first lemma shows how the usual variation splits at sets K of a given form, and the second gives the uniformity result.

(19.2) LEMMA. Let K be a finite union of closed intervals and let H be the complement of the interior of K. Then if S is filtering,
$$V(\psi, S; X) = V_K(\psi, S; X) + V_H(\psi, S; X).$$

PROOF. For any S-cover C of the set X let us write
$$C_1 = \{ I \in C : \text{either } I \subset K \text{ or } I \subset H \}.$$
It is easy to see that C_1 is also an S-cover of X and that
$$\text{Var}(\psi, C_1) = \text{Var}_K(\psi, C_1) + \text{Var}_H(\psi, C_1).$$

From this the assertion of the lemma may be deduced.

(19.3) LEMMA. Let S be filtering, let $\epsilon > 0$, and suppose that
$$V(\psi, S; X) < +\infty .$$
For any S-cover C of X, if the inequality
$$\text{Var}(\psi, C) \leq V(\psi, S; X) + \epsilon$$
holds, then for any set K that is a finite union of closed intervals,
$$\text{Var}_K(\psi, C) \leq V_K(\psi, S; X) + \epsilon .$$

PROOF. If we have the inequality
$$\text{Var}(\psi, C) \leq V(\psi, S; X) + \epsilon ,$$
if K is a finite union of intervals, and if H is the complement of the interior of K then, by lemma (19.2),
$$\text{Var}_K(\psi, C) \leq \text{Var}(\psi, C) - \text{Var}_H(\psi, C)$$
$$\leq V(\psi, S; X) + \epsilon - V_H(\psi, S; X) = V_K(\psi, S; X) + \epsilon$$
as required.

We now prove the statement of theorem (19.1). Let us suppose that $\{X_n\}$ is an increasing sequence of sets, which we may assume satisfies
$$V(\psi, S; X_n) < +\infty .$$
Let $\epsilon > 0$ and choose for each index n an S-cover C_n of X_n in such a way that
$$\text{Var}(\psi, C_n) \leq V(\psi, S; X_n) + \epsilon/2^n .$$
Construct the family of intervals C as
$$C = C_1 \cup (C_2 \setminus C_1) \cup (C_3 \setminus (C_1 \cup C_2)) \cdots$$
and note that C is an S-cover of the set X which is the union of the sets X_n. Let
$$\pi = \{I_1, I_2, I_3, \ldots, I_m\}$$
be any finite sequence of nonoverlapping intervals from C. We write
$$\pi_k = \{I_i : I_i \in C_k, I_i \notin C_j \text{ if } j < k\}$$
and let N be that index such that $\pi_k = \emptyset$ for $k > N$. Let K_k be the union of the intervals in π_k.

Now we compute, using the uniformity result from lemma (19.3) above,
$$\sum_{i=1}^{m} |\psi(I_i)| = \sum_{k=1}^{N} \sum_{I \in \pi_k} |\psi(I)|$$
$$\leq \sum_{k=1}^{N} \text{Var}_{K_k}(\psi, C_k)$$

$$\leq \sum_{k=1}^{N} \{ V_{K_k}(\psi, S; X_k) + \varepsilon/2^k \}$$

$$\leq \sum_{k=1}^{N} V_{K_k}(\psi, S; X_N) + \varepsilon$$

$$\leq V(\psi, S; X_N) + \varepsilon$$

$$\leq \sup \{ V(\psi, S; X_n) : n=1,2,3,\ldots \} + \varepsilon .$$

This estimate is available for any such choice of π from C and $\varepsilon > 0$ is arbitrary. Hence we obtain that

$$V(\psi, S; X) \leq \sup_n V(\psi, S; X_n)$$

as required to prove the increasing sets property.

Note that the increasing sets property is available for any <u>regular</u> outer measure (although it is not equivalent to the regularity of the measure). We can show that if ψ is a continuous interval function then the outer measure $S_{\bullet}\text{-}\psi$ is Borel regular, but may fail to be Borel regular if ψ is not continuous. Thus the question of the regularity of these measures appears to be somewhat delicate; for most applications the property proved in this section suffices and the regularity need not be checked.

The further investigation of these measures will be conducted in chapter four.

CHAPTER TWO

CLUSTER SETS

§20. Introduction.

In this chapter we begin a study of real cluster sets. It was W.H.Young [259], [260] who initiated the study of the cluster sets of an arbitrary real function (although the nomenclature is not due to Young). For an arbitrary function f, at a point x, one collects all extended real numbers y for which there is at least one sequence of numbers $\{x_n\}$ with $x_n \to x$ ($x_n \neq x$) and $f(x_n) \to y$. We denote this set of extended real numbers as $\Lambda(f, x)$ and refer to this as the cluster set of f at x.

Such a notion generalizes in several ways. We wish to retain the setting of the real line, but refine the notion of cluster set in a way that numerous authors have done before. Thus a variety of generalized cluster sets has been studied: right and left cluster sets, essential cluster sets (using density considerations), qualitative cluster sets (using category considerations), positive measure cluster sets, and so on. By using the notion of a local system of sets we obtain a generalization of these that permits us to formulate a great many results and, more importantly, to exhibit the methods rather clearly.

This chapter gives a review of the classical material on real cluster sets and develops some abstract presentations of this material. In particular, we will show how our standard machinery (intersection conditions, guage decomposition arguments, etc.) can be used to prove a number of theorems on real cluster sets. There are three basic types of theorems in this study; these concern a weak form of continuity, asymmetry theorems, and ambiguity theorems.

For further material on cluster sets see the survey article of C.Belna [10]. Other related questions appear in Ceder [47] and Rinne [205].

§21. S-cluster sets.

A number y (including $+\infty$ or $-\infty$) is said to be simply approached at a point x for a function f provided that there is at least one sequence of numbers $\{x_n\}$, distinct from x and converging to x, so that $f(x_n) \to y$. The set of all numbers so approached is denoted as $\Lambda(f, x)$ and is referred to as the cluster set of f at the point x.

A number of theorems can be easily established. Even for a completely arbitrary function f many statements can be made about the nature of these cluster sets. For example it is easy to see that each set $\Lambda(f, x)$ is nonempty and closed (as a subset of the extended real numbers). Collingwood [49] proved that the set of points

$$\{ x : f(x) \notin \Lambda(f, x) \}$$

at which an arbitrary function f does not have its value f(x) simply approached is at most denumerable.

If one refines the notion slightly so as to distinguish between values that are approached on one side, rather than on either side, further assertions are possible. Let $\Lambda^+(f, x)$ denote the set of limit numbers that are approached on the right, and $\Lambda^-(f, x)$ the set of limit numbers that are approached on the left. It is clear that the three sets $\Lambda(f, x)$, $\Lambda^+(f, x)$, and $\Lambda^-(f, x)$ are in general distinct; the only requirement that appears necessary is the obvious relation

$$\Lambda(f, x) = \Lambda^+(f, x) \cup \Lambda^-(f, x).$$

W.H.Young [260] obtained the interesting and remarkable result that the set of points where these three sets are not identical is denumerable, and this with no assumptions at all on the function f. Indeed Young claimed that this was the first true theorem that had been proved for a completely arbitrary real function.

These results have been subjected to considerable generalizations. In order to provide a systematic way of looking at these generalizations we consider the notion in the setting of a local system.

(21.1) DEFINITION. Let S be a local system and let f be an arbitrary function. Then by $(S)-\Lambda(f, x)$ we denote the set of all extended real numbers y with the property that for any neighbourhood V of y one has

$$f^{-1}(V) \cup \{x\} \in S(x).$$

Such a number is said to be S-<u>approached</u> by f at x and the collection of all S-approached numbers is the S-<u>cluster set</u> of f at x.

We list a variety of examples to illustrate the ideas as well as to provide references to the equivalent concepts that have appeared in the literature.

(21.2) Example. As usual take S_0 as the ordinary neighbourhood system, and S_∞ as its dual. Then it is easy to see that the S_∞-cluster set is the cluster set defined above, that is that

$$(S_\infty)-\Lambda(f, x) = \Lambda(f, x).$$

Note that the cluster set $(S_0)-\Lambda(f, x)$ can contain at most a single point c, and then only if $\lim_{y \to x} f(y) = c$.

(21.3) Example. Take S_0^+ and S_0^- as the right and left neighbourhood systems. Then the duals S_∞^+ and S_∞^- describe the usual right and left cluster sets,

$$(S_\infty^+)-\Lambda(f, x) = \Lambda^+(f, x) \quad \text{and} \quad (S_\infty^-)-\Lambda(f, x) = \Lambda^-(f, x).$$

(21.4) **Example.** In section 12 we have defined a number of local systems that used the notion of density. Corresponding to each of these is a type of cluster set. Here we give just the ones that have already been studied in the literature.

(i) The largest of the density families is the system S^1 described so that each $S^1(x)$ contains every set S for which $x \in S$ and such that S has exterior upper density positive on one side at least at x. The corresponding cluster set, denoted as
$$\Lambda_{ess}(f,x),$$
is known as the <u>essential cluster set</u> in the literature. Note that this set contains all extended real numbers y such that $f^{-1}(U_y)$ does not have x as a point of dispersion for any neighbourhood U_y of y. One-sided versions of this we will write as
$$\Lambda_{ess}^+(f,x) \quad \text{and} \quad \Lambda_{ess}^-(f,x).$$

(ii) The stronger density system S^2 described by requiring instead exterior, upper density 1 was used by Zajíček [268] in his study of cluster sets. Belna [10] denotes this cluster set as,
$$(S^2)-\Lambda(f,x) = H(f,x).$$

(iii) The density system S^3 using exterior (lower) density 1 was introduced into cluster set language by Belna [10] using the notation,
$$(S^3)-\Lambda(f,x) = HB(f,x).$$

All of the other density systems can used in this context but there does not appear to have been any such general study in the literature.

(21.5) **Example.** A system based on sets of positive measure may be used in this setting. Define S^4 so that each $S^4(x)$ contains sets that have exterior positive measure in every right or left neighbourhood of the point x. Following Belna [10] we use the notation
$$(S^4)-\Lambda(f,x) = M(f,x).$$
This is studied in Zajíček [268].

(21.6) **Example.** (Topological setting). The cluster set notion may be generalized by using some elementary ideas from topology. If τ is a topology on the real line then it is natural to define the τ-cluster set, $\Lambda_\tau(f,x)$, of an arbitrary function f at a point x as the set
$$\{y : x \text{ is a } \tau\text{-accumulation point of } f^{-1}(V), \text{ all } V \text{ n'ds of } y\}.$$
Hunter [125] takes this idea and obtains a general asymmetry theorem for cluster sets relative to two different topologies.

There is an obvious choice of system so that this cluster set will fit into the present framework.

(21.7) Example. (Closure operator). Zajíček [268] carries the previous example over to the setting of a closure operator (that is not necessarily the closure operator for an actual topology). If u is a closure operator (see section 10 of Chapter one) then define
$$\Lambda_u(f, x) = \{ y : x \in der_u(f^{-1}(V)) \text{ for every n'd } V \text{ of } y \}.$$
where $der_u(A)$ is the "derivative" of the set relative to the closure operator u as has been previously defined.

There is a local system S_u so that the cluster sets
$$(S_u)-\Lambda(f, x) = \Lambda_u(f, x)$$
are identical. Recall that this formulation of Zajíček using a closure operator has been shown in lemma (10.1) to be equivalent to the use of a local system that has a filtering dual.

(21.8) Example. Let **N** be a σ-ideal of sets of real numbers. We say that an extended real number c is <u>weakly **N**-approached</u> on the right by a function f at a point x, provided that for every neighbourhood V of c,
$$f^{-1}(V) \cap (x, x + \delta)$$
is not in **N** for every $\delta > 0$. The set of such numbers is denoted as $\Lambda_N^+(f, x)$ and referred to as the right cluster set relative to the σ-ideal. Similarly we define a left hand version and a bilateral version. Clearly these are cluster sets relative to suitable local systems.

§22. Elementary theorems for cluster sets.

In this section we collect several immediate properties of such cluster sets. With the assumption of a variety of conditions on the local system S there are many results available and an extensive theory is possible. The first task is to establish the basic properties.

The simplest result we can give in this setting offers a comparison between the two cluster sets $(S_1)-\Lambda(f, x)$ and $(S_2)-\Lambda(f, x)$ relative to two local systems S_1 and S_2. We can consider such an assertion as giving a relation between the cluster sets relative to a pair of systems. Assertions of this type are central to our concerns in this study; the relation here is the simplest and requires the least machinery. More complicated relations are studied in later sections.

(22.1) LEMMA. Let S_1 and S_2 be local systems for which $S_1 \ll S_2$. Then one has for any function f, and at any point x,
$$(S_1)-\Lambda(f, x) \subset (S_2)-\Lambda(f, x).$$
Conversely, if this relation holds for these cluster sets at any point x and for all functions f, then $S_1 \ll S_2$.

PROOF. If y is a member of the cluster set (S_1)-$\Lambda(f,x)$ then, for any neighbourhood U of y, one has that $f^{-1}(U)$ belongs to $S_1(x)$. Since $S_1(x) \subset S_2(x)$ this means that $f^{-1}(U)$ is in $S_2(x)$ for every such U and, by definition, then y is in (S_2)-$\Lambda(f,x)$. This proves the first part of the lemma.

For the second part, if the order relation $S_1 \ll S_2$ does not hold, then there must be a point x and a set S that belongs to $S_1(x)$ but not to $S_2(x)$. We take for f the characteristic function of this set S. Then, because
$$S = \{y : |f(y) - 1| < \epsilon\}$$
for every $1 > \epsilon > 0$, it is clear that 1 is in the cluster set (S_1)-$\Lambda(f,x)$. But, since S does not belong to $S_2(x)$, we have that 1 is not in the cluster set (S_2)-$\Lambda(f,x)$. This proves the second part.

Since every local system S lies between the two extremes S_0 and S_∞, it must be the case that the cluster sets relative to an arbitrary system S satisfy
$$(S_0)\text{-}\Lambda(f,x) \subset (S)\text{-}\Lambda(f,x) \subset (S_\infty)\text{-}\Lambda(f,x).$$
This says merely that (S)-$\Lambda(f,x)$ is always a subset of the ordinary cluster set $\Lambda(f,x)$ and that, should the function have a genuine limit at a point x, then (S)-$\Lambda(f,x)$ can contain only that limit.

For an arbitrary function f it is easy to show that the basic cluster set $\Lambda(f,x)$ must be a nonempty closed set in the extended reals. If S is an arbitrary local system then the generalized cluster set (S)-$\Lambda(f,x)$ is closed, but may well be empty. The example $S = S_0$ easily illustrates this fact, for this cluster set will be empty unless the actual limit of the function exists at the point. Thus in order to obtain a general theorem asserting that the cluster set S-$\Lambda(f,x)$ is nonempty we shall need a mild hypothesis on the system S.

(22.2) LEMMA. Let S be a local system of sets; then every cluster set (S)-$\Lambda(f,x)$ is closed. In order that it also be nonempty for all functions f it is sufficient that S have a filtering dual at x (i.e that S has the property $[J_1]$ of section 5).

PROOF. For an arbitrary system S the cluster set (S)-$\Lambda(f,x)$ must be closed (although it may be empty). Suppose that y is an accumulation point of the set (S)-$\Lambda(f,x)$. To keep the arguments simple we will assume y is finite (note that the closure is being taken in the extended real number system). Then, for any $\epsilon > 0$, there is a point $y_1 \in (S)$-$\Lambda(f,x)$ in the interval $(y - \epsilon/2, y + \epsilon/2)$. The set
$$S = \{t : t = x \text{ or } |f(t) - y| < \epsilon\}$$
must contain the set
$$S_1 = \{t : t = x \text{ or } |f(t) - y_1| < \frac{\epsilon}{2}\}$$

and this set belongs to $S(x)$. Thus $S \in S(x)$ for any $\varepsilon > 0$ and this shows that y is a member of the cluster set $(S)-\Lambda(f, x)$.

It remains now to show that, under the hypothesis $[J_1]$ on S at the point x, the cluster set (as a set of extended real numbers) must be nonempty. If it is empty then, for every extended real number y, there is a neigbourhood U_y of y so that
$$\{x\} \cup f^{-1}(U_y) \not\in S(x). \qquad (*)$$
By an elementary compactness argument there is a sequence of numbers $\{y_1, y_2, \ldots, y_n\}$ so that the sets
$$U_{y_i} \qquad i = 1, 2, 3, \ldots, n$$
cover the extended real line. Then certainly
$$\sum_{i=1}^{n} f^{-1}(U_{y_i}) = \mathbb{R} \in S(x)$$
which together with the hypothesis $[J_1]$ and $(*)$ above is a contradiction.

From this proof we may extract a result that will be useful to us on occasion. This is only a refinement of the lemma.

(22.3) LEMMA. Let S have filtering dual, let f be a real function and C a set of real numbers for which
$$f^{-1}(C) \cup \{x\} \in S(x) .$$
Then there must exist in the closure of C (the closure of C in the extended real line) a value of $(S)-\Lambda(f, x)$.

As a further consequence we may derive the following useful observation.

(22.4) LEMMA. Let S have a filtering dual. Then for any open set $G \subset \overline{\mathbb{R}}$, if
$$G \supset (S)-\Lambda(f, x)$$
then necessarily
$$f^{-1}(G) \cup \{x\} \in S^*(x) .$$
Conversely if this latter condition holds then
$$\overline{G} \supset (S)-\Lambda(f, x) .$$

PROOF. Let the open set G contain the cluster set and let K be the complement of G in $\overline{\mathbb{R}}$. Then by the preceding lemma the set
$$f^{-1}(K) \cup \{x\}$$
cannot belong to $S(x)$, otherwise there would be a cluster set value in K. Consequently
$$f^{-1}(G) \cup \{x\} = [\mathbb{R} \setminus f^{-1}(K)] \cup \{x\} \in S^*(x)$$
as required.

Conversely suppose that $f^{-1}(G) \cup \{x\} \in S^*(x)$ and let U be the complement of the closure of G in $\overline{\mathbb{R}}$. Then, if there is a cluster set value at x outside of $\overline{G}$,

$$[\mathbb{R} \setminus f^{-1}(G)] \cup \{x\} \supset f^{-1}(U) \cup \{x\} \in S(x).$$

But this is impossible and the lemma is proved.

Finally, to conclude these concerns, we might ask whether the condition $[J_1]$ is both necessary and sufficient in order that the cluster sets relative to a system be nonempty. This condition $[J_1]$ is equivalent, by lemma (5.1), to the requirement that S^* should be filtering. It is not difficult to construct an example to show that this is not the case; the condition $S^* \ll S$ is however necessary. The condition that may be proved to be both necessary and sufficient is the following variant of $[J_1]$ used in the next lemma.

(22.5) LEMMA. In order that the system S have the property that every cluster set $(S)-\Lambda(f, x)$ is nonempty, it is necessary and sufficient that S have the property that, whenever

$$A_1, A_2, \ldots A_n$$

is a finite sequence of sets covering the real line, one at least of the sets

$$A_i \cup \{x\}$$

must belong to $S(x)$.

PROOF. The proof in one direction is similar to the proof of (22.2) above. In the converse direction, suppose that there is such a sequence $\{A_i\}$ but that none of the sets

$$A_i \cup \{x\}$$

belongs to $S(x)$. Then define the function f to be 1 on A_1, 2 on $A_2 \setminus A_1$ and so on. Then $(S)-\Lambda(f, x) = \emptyset$. This shows the condition is necessary and the proof is complete.

§23. S-derived set.

We introduce in this section a concept which is of frequent use in the sequel. The notion is topological in nature and closely related to the ideas of "neighbourhood" and "accumulation point", but expressed in the language of local systems. We use the notation S-der [A] to suggest that a derived set (in the topological sense) is considered.

(23.1) DEFINITION. Let S be a local system and let A be an arbitrary set. Then by the S-<u>derived set</u> of A we mean the set of points

$$(S) - \text{der}[A] = \{x : A \cup \{x\} \in S(x)\}.$$

We have a closely related definition.

(23.2) DEFINITION. Let S be a local system of sets on a set X. A set N will be said to be (S)-<u>negligible</u> provided that for each $x \in N$ the set N does not belong to $S(x)$. A set is said to be σ-(S)-<u>negligible</u> if it can be expressed as a countable union of sets that are (S)-negligible.

Observe that a set A is (S)-negligible if and only if
$$A \cap (S)\text{-der}[A] = \emptyset.$$
We may express this dually as well. A set A is (S)-negligible if and only if A is a subset of
$$(S^*)\text{-der}[\mathbb{R} \setminus A].$$
The class of σ-(S)-negligible sets is clearly a σ-ideal that contains the denumerable sets.

The examples illustrate the notions of (S)-derived set and of (S)-negligible set.

(23.3) Example. For the extreme examples S_0 and S_∞ one can verify that the derived set (S_0)-der$[A]$ is precisely the set of points
$$\{ x : x \text{ is an interior point of } A \cup \{x\} \}.$$
Similarly (S_∞)-der$[A]$ is the set of points
$$\{ x : x \text{ is a point of accumulation of } A \}.$$
Consequently a set A is (S_0)-negligible if and only if it has no interior points, and a set A is (S_∞)-negligible if and only if it contains only isolated points. From this we may conclude that any set of real numbers is (S_0)-σ-negligible, and that it is the denumerable sets that are (S_∞)-σ-negligible.

(23.4) Example. As in example (21.4) let S^3 denote the system described by requiring that $S \in S^3(x)$ if and only if $x \in S$ and S has lower exterior density 1 on both sides at x. By the density theorem if A is any set then, for almost every point $x \in A$, the exterior density of A at x is 1. From this we deduce that $A \setminus (S^3)$-der$[A]$ has measure zero. Thus the class of (S^3)-negligible sets is just the class of measure zero sets.

(23.5) Example. Let N be a σ-ideal of sets that contains no interval and let S be the local system described as follows: $S \in S(x)$ if and only if $x \in S$ and for no positive number δ do both of the sets
$$S \cap (x, x+\delta) \quad \text{and} \quad S \cap (x-\delta, x)$$
belong to the σ-ideal N.

Let us check that a set E is σ-(S)-negligible if and only if E is the union of a denumerable set and a set that belongs itself to the σ-ideal N.

Certainly if $E \in N$ or if E is finite then E is (S)-negligible. For the converse suppose that E is (S)-negligible. There is a gauge δ on E so that one of the sets

$$E \cap (x, x + \delta(x)) \quad \text{and} \quad E \cap (x - \delta(x), x)$$

belongs to **N**. We may suppose that at every $x \in E$ it is the former set that belongs to **N**. Let $\{E_n\}$ denote the guage partition of E induced by δ. It is easy to check that each E_n is the union of a finite set and a set in the σ-ideal **N**.

We need, on occasion, to know the Borel structure of an S-derived set and we continue this section with several results of this type for specific systems.

(23.6) **Example.** For S_0, the ordinary neighbourhood system, the set of points
$$(S_0) - \text{der}[A]$$
is the set of all points x for which $A \cup \{x\}$ is a neighbourhood of x. Consequently this derived set is easily seen to be open.

(23.7) **Example.** The derived set
$$(S_\infty) - \text{der}[A]$$
is precisely the set of points of accumulation of A and so is clearly closed.

(23.8) **Example.** The set of points
$$(S_\infty^+) - \text{der}[A]$$
is the set of points of right hand accumulation of A. This can be seen to be a difference of a closed set and a denumerable set; in particular it is a G_δ.

(23.9) **Example.** We wish now to establish the Borel classification of the derived sets for the density systems. We pause to prove an elementary result that is needed:

(23.10) **Lemma.** Let A be an arbitrary set and define the real function g by writing
$$g(x) = \limsup_{t \to 0+} \frac{|A \cap (x, x+t)|^e}{t}.$$
Then $g(x)$ computes the upper, exterior right hand density of A at any point x and g is in Baire class 2.

For a proof (cf. Zajíček [268, p.206]) we define the function h of two variables,
$$h(x, r) = |A \cap (x, x + r)|^e / r,$$
which, for fixed $r > 0$, is evidently continuous. We observe that the value $g(x)$ may be expressed as
$$\lim_{m \to \infty} \lim_{j \to \infty} \max\{g(x, r) : 1/j \leq r \leq 1/m\}$$
and we see from this that g must be at least Baire class 2.

By the result just proved the set of points,
$$(S^+_{ess})-\text{der}[A] = \{x : g(x) > 0\},$$

for any set A, must be of type $G_{\delta\sigma}$, where S^+_{ess} has been defined above in example (21.4). A similar statement may be made for the left derived set (S^-_{ess})-der $[A]$.

§24. Weak continuity theorems.

An arbitrary function need have no points of continuity. Nonetheless every function possesses a number of weak continuity properties that hold at most points. For example, let us say that a function f is <u>feebly continuous</u> at a point x if there is at least one sequence of numbers x_n with $x_n \neq x$, $x_n \to x$, and $f(x_n) \to f(x)$. Equivalently this says that the number $f(x)$ belongs to the cluster set $\Lambda(f,x)$. W.H. Young [259] showed that, with at most denumerably many exceptions, the values $f(x)$ lie between the extremes of the cluster set $\Lambda(f,x)$; Collingwood later showed that, for an arbitrary function f, the membership
$$f(x) \in \Lambda(f,x)$$
must hold for all x except at a denumerable set. Thus an arbitrary function is feebly continuous at every point with at most denumerably many exceptions.

Since then there have been a number of theorems of this type asserting that an arbitrary function has some kind of weak continuity property except at some relatively small set. Expressed in our language the problem becomes to characterize the set of points
$$\{ x : f(x) \notin (S)\text{-}\Lambda(f,x) \}$$
for a given local system S.

The theorem below contains the result of Collingwood as well as a related theorem of Blumberg (see example (24.3) below), and several others; it is just an elementary expression of the general idea used in Collingwood's theorem. Roughly it asserts that every function has a weak form of continuity at "most" points.

(24.1) THEOREM. Let S be a local system of sets and f an arbitrary function. Then the set of points
$$\{ x : f(x) \notin (S)\text{-}\Lambda(f,x) \}$$
must be σ-(S)-negligible.

PROOF. Let X be the exceptional set of the theorem,
$$X = \{ x : f(x) \notin (S)\text{-}\Lambda(f,x) \},$$
and choose for each point $x \in X$ a number $\nu(x) \in (0,1)$ so that each set
$$S_x = \{ t : t = x \text{ or } |f(t) - f(x)| < \nu(x) \}$$
fails to belong to $S(x)$. Let $\{X_n\}$ be the partition of X induced by the guage ν on X, by writing
$$X_n = \{ t \in X : \frac{1}{n} < \nu(x) \leq \frac{1}{n-1} \}.$$

Partition each X_n further by writing

$$X_{nm} = \{ x \in X_n : f(x) \in [\tfrac{m}{2n}, \tfrac{m+1}{2n}) \},$$

for $m = \ldots, -3, -2, -1, 0, 1, \ldots$ Notice that, for any pair of points x and y in the same set X_{nm}, the values $f(x)$ and $f(y)$ are within $1/2n$ of each other, i.e. that

$$|f(x) - f(y)| < \tfrac{1}{2n} < \tfrac{1}{n} < \nu(x).$$

From this we see that $X_{nm} \subset S_x$ and so X_{nm} cannot belong to $S(x)$ for any point $x \in X$. This means that X_{nm} is (S)-negligible, and consequently that X itself is σ-(S)-negligible as required.

The examples illustrate the theorem.

(24.2) Example. Take S_0 as the usual neighbourhood filter at each point; any set is σ-(S_0)-negligible and so the theorem applied to this system is vacuous. For the dual S_∞ we can easily check that a set is (S_∞)-negligible if and only if it contains no point of accumulation; thus a σ-(S_∞)-negligible set is denumerable and this gives us the theorem of Collingwood stated above: for any function f the set of points

$$\{ x : f(x) \not\in \Lambda(f, x) \}$$

is denumerable.

(24.3) Example. Applied to the density system S^3 of examples (21.4) and (23.4) we have seen that any S^3-negigible set must have measure zero. Thus we obtain (using the language of (21.4)(iii)) for an arbitrary function f, that

$$f(x) \in HB(f, x)$$

for almost every point x. Occasionally this is expressed in the literature as asserting that <u>every</u> function is "weakly approximately continuous" almost everywhere. Equivalently this asserts that at almost every point x there is a set E_x having exterior density 1 at x and such that

$$\lim_{y \in E_x, y \to x} f(y) = f(x).$$

(24.4) Example. Let N be a σ-ideal of sets that contains no interval and let S be the local system described as follows: $S \in S(x)$ if and only if $x \in S$ and for no positive number δ does the set $S \cap (x, x+\delta)$ belong to the σ-ideal N. We know from example (23.5) that an σ-(S)-negligible set is the union of a denumerable set and a set that belongs to the σ-ideal. Thus we may apply the basic theorem to obtain that every function is "weakly N-continuous" (on both sides) except at the points of such a set. By weak N-continuity on the right at a point x_0 we mean that the set

$$\{ y : |f(y) - f(x_0)| < \epsilon \} \cap (x_0, x_0 + \epsilon)$$

does not belong to N for any $\epsilon > 0$. This may be expressed as asserting that there is a set

E_x for which $E_x \cap (x, x+\varepsilon)$ does not belong to **N** for any $\varepsilon > 0$ and such that
$$\lim\nolimits_{y \,\in\, E_x,\, y \,\to\, x} f(y) = f(x).$$

In particular we would obtain the following results as special cases: for any function f there are at most denumerably many points x at which both of the sets
$$\{y : |f(y) - f(x)| < \delta\} \cap (x, x+\delta)$$
$$\{y : |f(y) - f(x)| < \delta\} \cap (x-\delta, x)$$
are not nondenumerable for all $\delta > 0$. Similarly there is a set of the first category so that, for all x not in that set, each of the sets above is second category for all $\delta > 0$. Again there is a set of measure zero so that, for all x not in that set, each of the sets above has positive exterior measure.

(24.5) Example. Our theorem (24.1) can be used to show that for an arbitrary function f there is a σ-porous set N such that every set
$$\{t : a < f(t) < b\}$$
is nonporous at each of its points, except possibly those that are in N. This is related to the M_3-property of Zahorski [265].

Finally let us conclude by restating the main theorem of this section but in two alternative formulations which may be of use in certain applications. In the first formulation we use the fact that many applications have a σ-ideal of exceptional sets given *a priori*. In that case one would not investigate the class of S-negligible sets, but immediately verify that the sets in the σ-ideal are the needed exceptional sets.

(24.6) THEOREM. Let **N** be a σ-ideal of sets. Suppose that the system **S** has the property that, if a set A has the property that $A \notin S(x)$ for every point $x \in A$, then necessarily $A \in \mathbf{N}$. Then for every function f the set,
$$\{x : f(x) \notin (S)\text{-}\Lambda(f, x)\},$$
belongs to **N**.

A further alternative version reveals the exact structure of the exceptional set.

(24.7) THEOREM. For an arbitrary system **S** and an arbitrary function f the set of points
$$\{x : f(x) \notin (S)\text{-}\Lambda(f, x)\}$$
may be displayed in the form
$$\sum_{n=1}^{\infty} \left(A_n \setminus (S)\text{-der}\,[A_n] \right)$$

for a sequence of sets A_n.

PROOF. This is proved similarly to the theorem but put $A_n = f^{-1}(U_n)$ where U_n ranges over a base for the neighbourhoods in the extended real line.

§25. Asymmetry theorems.

It is clear that at any particular point, for an arbitrary function f, the two one-sided cluster sets $\Lambda^+(f, x)$ and $\Lambda^-(f, x)$ need have no points in common. W.H.Young [260] investigated the relation between these two sets and observed that, except at a denumerable set, these two cluster sets must be intimately connected. He first obtained a relation between the upper and lower bounds of the pair, and then improved this to obtain the remarkable conclusion that, for a completely arbitrary function f, the two sets $\Lambda^+(f, x)$ and $\Lambda^-(f, x)$ are identical except at a denumerable exceptional set. As he presented this theorem at the 1908 international congress at Rome, he referred to this result as his "Rome theorem".

In an abstract setting the distinction of right and left disappears and so a generalization of this theorem will take the form of investigating the relation, if any, that must hold between the two cluster sets (S_1)-$\Lambda(f, x)$ and (S_2)-$\Lambda(f, x)$ for a given pair of systems S_1 and S_2. We shall refer to this investigation under the heading of an **asymmetry** relation.

Hunter [125] developed a general approach to this problem by considering that there have been given two topologies τ_1 and τ_2 on the real line; he then posed the problem of determining some topological way of describing the set

$$\{ x : \Lambda_{\tau_1}(f, x) \neq \Lambda_{\tau_2}(f, x) \}$$

of asymmetry points for a function relative to these two topologies. (See example (21.6) for the notation that is used here.) Zajíček [268] has refined these ideas of Hunter and extended them to the more general setting of a closure operator, as in example (21.7). Related notions can be found in the works of Świątkowski [226] and Jędrzejewski [129]; these are discussed in example (26.4).

In this section we place these ideas in our setting and develop them somewhat further. The key idea in Hunter's approach was to determine the nature of the set of points that are accumulation points of a given set relative to one of the two topologies but not to both. This idea can be expressed in the following definition where we use two local systems S_1 and S_2 rather than two topologies.

(25.1) DEFINITION. Let S_1 and S_2 be local systems, and let A be a set of real numbers. Then we define the sets

$$(S_1 \setminus S_2)\text{-der}[A] = (S_1)\text{-der}[A] \setminus (S_2)\text{-der}[A]$$

and

$$(S_1 \triangle S_2)\text{-der}[A] = (S_1)\text{-der}[A] \triangle (S_2)\text{-der}[A]$$

Here "$\setminus$" denotes the usual set difference but in the expression $S_1 \setminus S_2$ it is used only symbolically; similarly "$\triangle$" denotes the symmetric difference

$$C_1 \triangle C_2 = (C_1 \setminus C_2) \cup (C_2 \setminus C_1) ,$$

but we do not interpret the symbol $S_1 \triangle S_2$. Note that these notions are just shorthand for the set of points that are in the derived set for one system but not for the other. Thus we have in fact for any pair of systems S_1 and S_2 that the expression

$$(S_1 \setminus S_2)\text{-der}[A]$$

denotes the set

$$\{ x \in \mathbb{R} : A \cup \{x\} \in S_1(x) \setminus S_2(x) \} ,$$

and, similarly, that

$$(S_1 \triangle S_2)\text{-der}[A]$$

denotes

$$\{ x \in \mathbb{R} : A \cup \{x\} \in S_1(x) \triangle S_2(x) \} .$$

The next two lemmas show how this concept can be exploited to give a representation of an asymmetry set. The first of these is basically Hunter's idea and the second, which gives the more precise representation, is an improvement in the spirit of Zajíček [268].

(25.2) LEMMA. Let S_1 and S_2 be two local systems and f an arbitrary function. Then the set of points where one cluster set does not include the other,

$$\{ x : (S_1)\text{-}\Lambda(f, x) \setminus (S_2)\text{-}\Lambda(f, x) \neq \emptyset \} ,$$

is a subset of a set of the form

$$\sum_{n=1}^{\infty} (S_1 \setminus S_2) \text{-der}[A_n]$$

for some sequence of sets A_n.

PROOF. We take $\{U_n\}$ as a base for the open sets of $\overline{\mathbb{R}}$, and write $A_n = f^{-1}(U_n)$. It remains to show that, with this choice of the sets A_n, the stated set inequality does indeed hold. If x is a member of the first stated set, then there must be a number $g(x)$ ($-\infty \leq g(x) \leq +\infty$), so that

$$g(x) \in (S_1)\text{-}\Lambda(f, x) \quad \text{and} \quad g(x) \notin (S_2)\text{-}\Lambda(f, x) .$$

Take a neighbourhood U_m of $g(x)$ so that $f^{-1}(U_m) \cup \{x\}$ is not in $S_2(x)$. Then with

$A_m = f^{-1}(U_m)$ as before we must have
$$A_m \cup \{x\} \in S_1(x) \text{ but } A_m \cup \{x\} \notin S_2(x).$$
Thus x is a member of the set $(S_1 \setminus S_2)$-der$[A_m]$ as required. As this may be done for any point x in the set on the left, the asserted inequality is proved.

(25.3) LEMMA. Let S_1 and S_2 be two local systems such that S_1 has a filtering dual and let f be an arbitrary function. Then the set of points where one cluster set does not include the other,
$$\{ x : (S_1)\text{-}\Lambda(f, x) \setminus (S_2)\text{-}\Lambda(f, x) \neq \emptyset \},$$
can be exactly represented as

$$\sum_{n=1}^{\infty} (S_1 \setminus S_2)\text{-der}[A_n] \cap (S_1 \setminus S_2)\text{-der}[B_n]$$

for some sequences of sets $\{A_n\}$ and $\{B_n\}$.

PROOF. From a countable base for the open sets of $\overline{\mathbb{R}}$ we select pairs (U_n, V_{nm}) so that U_n ranges over the the base and V_{nm} ranges over all elements in the base for which $\overline{V_{nm}} \subset U_n$. We may consider that this has been relabelled and indexed as a single sequence $\{(U_k, V_k)\}$ ($k = 1, 2, 3, \ldots$) and we write $A_k = f^{-1}(U_k)$ and $B_k = f^{-1}(V_k)$. It remains to show that this indeed gives the required representation.

By the preceding lemma we certainly know that there is set containment, and so it is the set equality that we require. Suppose then that x is a point that belongs to both of the sets
$$(S_1 \setminus S_2)\text{-der}[A_n] \text{ and } (S_1 \setminus S_2)\text{-der}[B_n]$$
for some index n. Then this must mean that
$$f^{-1}(U_n) \cup \{x\} \in S_1(x), \quad f^{-1}(U_n) \cup \{x\} \notin S_2(x),$$
$$f^{-1}(V_n) \cup \{x\} \in S_1(x), \text{ and } f^{-1}(V_n) \cup \{x\} \notin S_2(x).$$

If we interpret these assertions now, remembering that the closure of V_n is contained in U_n, we see (because of the fact that S_1 has a filtering dual and from lemma (22.3)) that at the point x there must be a cluster set value $g_1(x)$ in (S_1)-$\Lambda(f, x)$ that lies inside the closure of V_n and yet there are no cluster set values $g_2(x)$ in (S_2)-$\Lambda(f, x)$ that lie within the set U_n. Consequently x must lie in the set on the left of the equality as required. Since this argument applies to any point x in the right hand side of the equation, the lemma is proved.

Each of the two lemmas above may be given a version in which inequality rather than a failed set inclusion is asserted. With essentially the same proofs we state the following lemmas.

(25.4) LEMMA. Let S_1 and S_2 be two local systems and let f be an arbitrary function. Then the set of points where the two cluster sets are not equal,
$$\{ x : (S_1)\text{-}\Lambda(f, x) \neq (S_2)\text{-}\Lambda(f, x) \},$$
is a subset of a set of the form
$$\sum_{n=1}^{\infty} (S_1 \Delta S_2)\text{-der}[A_n]$$
for some sequence of sets $\{A_n\}$.

(25.5) LEMMA. Let S_1 and S_2 be two local systems each with a filtering dual and let f be an arbitrary function. Then the set of points where the two cluster sets are not equal,
$$\{ x : (S_1)\text{-}\Lambda(f, x) \neq (S_2)\text{-}\Lambda(f, x) \}$$
may be precisely represented as
$$\sum_{n=1}^{\infty} (S_1 \Delta S_2)\text{-der}[A_n] \cap (S_1 \Delta S_2)\text{-der}[B_n]$$
for some sequences of sets $\{A_n\}$ and $\{B_n\}$.

Before concluding our discussion of these general asymmetry theorems let us present an alternative version (in the same spirit as Theorem (24.6) above). In certain instances one wishes to know immediately whether the exceptional set belongs to some specified σ-ideal. With essentially the same proofs as above, we assert the following theorem, using the appropriate definition.

(25.6) DEFINITION. Let us say that, for a pair of systems S_1 and S_2, a joint intersection condition <u>fails</u> on the set X if for every $x \in X$ one may choose sets
$$S^1_x \in S_1(x) \quad \text{and} \quad S^2_x \in S_2(x)$$
so that the intersection $S^1_x \cap S^2_y$ contains no points other than possibly x or y for each pair of points x and y in X.

(25.7) THEOREM. Let S_1 and S_2 be local systems and N a σ-ideal of sets. Suppose that it is true that whenever a joint intersection fails on a set X for the pair of systems S_1 and S_2^* it must be the case that $X \in N$. Then for any function f the set
$$\{ x : (S_1)\text{-}\Lambda(f, x) \setminus (S_2)\text{-}\Lambda(f, x) \neq \emptyset \}$$
belongs to N.

§26. Young's Rome theorem.

In this section we shall apply the results of the preceding section to obtain a number of versions of Young's theorem, that is to say a number of asymmetry theorems in which the exceptional set is denumerable. Our first asymmetry theorem simply exploits an intersection condition. This theorem will include the original Rome theorem as a special case.

(26.1) DEFINITION. Let S_1 and S_2 be a pair of local systems on a set X. We say that the pair (S_1, S_2) satisfies a <u>joint intersection condition</u> if for any choices
$$\{S_1^x : x \in X\} \quad \text{and} \quad \{S_2^x : x \in X\}$$
with each $S_1^x \in S_1(x)$ and $S_2^x \in S_2(x)$ there is a gauge δ on X so that whenever
$$0 < |y - x| < \min\{\delta(x), \delta(y)\}$$
one at least of the intersections
$$S_1^x \cap S_2^y \quad \text{or} \quad S_2^x \cap S_1^y$$
contains points other than x and y.

(26.2) THEOREM. Let S_1 and S_2 be local systems and f an arbitrary function. If the pair (S_1^*, S_2) has the joint intersection condition expressed in the definition above, then the relation
$$(S_1)\text{-}\Lambda(f, x) \supset (S_2)\text{-}\Lambda(f, x)$$
must hold for all x except possibly at a denumerable set.

PROOF. This may be proved directly. Alternatively we may use lemma (25.2). For any set A we must show that the set
$$X = (S_2 \setminus S_1)\text{-der}[A]$$
is denumerable. For each $x \in X$ we write
$$S_x = A \cup \{x\} \quad \text{and} \quad T_x = (\mathbb{R} \setminus A) \cup \{x\}.$$
By our hypotheses we have $S_x \in S_2(x)$ and $T_x \in S_1^*(x)$ for each point x in X.

Corresponding to these choices from S_2 and S_1^* there is a gauge δ on X satisfying the conditions of the joint intersection definition; use this to induce a decomposition of the set X into a sequence $\{X_n\}$ that has the property that if there is a pair of points x, $y \in X_n$, then one of the intersections
$$S_x \cap T_y \quad \text{or} \quad T_x \cap S_y$$
must contain points other than x and y. As this is impossible we see that each set X_n can contain no more than a single point and consequently the set X must be denumerable as required to prove the theorem.

(26.3) **Example.** This theorem includes as a particular case the Rome theorem of Young. For the systems S_∞^+ and S_∞^- the asymmetry set
$$\{ x : (S_\infty^+)\text{-}\Lambda(f,x) \neq (S_\infty^-)\text{-}\Lambda(f,x) \}$$
or rather, in our earlier terminology,
$$\{ x : \Lambda^+(f,x) \neq \Lambda^-(f,x) \},$$
is denumerable. One may check that the required intersection conditions hold. Alternatively, using lemma (25.4) one checks easily that for any set A,
$$(S_\infty^+ \Delta\ S_\infty^-)\text{-der}[A]$$
consists of all points x such that x is a point of accumulation of A on one side only. Such a set is of course denumerable and so Young's theorem follows.

(26.4) **Example.** [Świątkowski] In Świątkowski [226] ideas of this type are developed in such a way as to obtain conditions on a topology τ, so that for any function f, the set of points of left-right assymmetry
$$\{ x : \Lambda_\tau^+(f,x) \neq \Lambda_\tau^-(f,x) \}$$
is denumerable. The condition he uses is the following.

(W) For any sequence of points $\{x_n\}$ converging to a point x and any sequence of sets $\{E_n\}$ if $x_n \in \text{der}_\tau[E_n]$ for each n then
$$x \in \text{der}_\tau\left[\sum_{n=1}^{\infty} E_n\right].$$
This same idea was explored by Jędrzejewski [129] in a setting closely related to our own. The following theorem contains the essence of both approaches.

(26.5) **Theorem.** Let S^+ and S^- be right and left hand simple systems, respectively. Suppose that $S = S^+ \vee S^-$ and that S has the property (W), namely that for every sequence of numbers $\{x_n\}$ converging to a point x and for every sequence of sets $\{E_n\}$, if each set $E_n \cup \{x_n\}$ belongs to $S(x_n)$ then the set
$$\{x\} \cup \left(\sum_{n=1}^{\infty} E_n\right)$$
belongs to $S(x)$. Then for any function f the set of right and left asymmetry
$$\{ x : (S^+)\text{-}\Lambda(f,x) \neq (S^-)\text{-}\Lambda(f,x) \}$$
is denumerable.

Proof. One has only to check that the joint intersection condition of the definition is available. The proof reduces to showing that, for any set A, the set of points
$$X = (S^+ \setminus S^-)\text{-der}[A]$$
is denumerable. If not then there is an increasing sequence of numbers $\{x_n\}$ in X con-

verging to a point x in X and each set $A \cup \{x_n\}$ belongs to $S^+(x_n)$ but $A \cup \{x\}$ does not belong to $S^-(x)$. By condition (W) this is impossible and so the theorem must be true.

(26.6) **Example.** Our general asymmetry theorem (26.2) uses a joint intersection condition on the pair (S_1, S_2^*). One might ask whether this condition is sharp. That is, if it is the case that for every function f the set of points
$$(S_1) - \Lambda(f, x) \neq (S_2) - \Lambda(f, x)$$
is denumerable, is it the case that the pair must satisfy this intersection condition? We give an example to show that this is not necessarily so.

Let C be a perfect nowhere dense set. Define the local system S_1 so that $S_1(x)$ is just the ordinary right neighbourhood system at x if x is not in C or is isolated on one side in C, otherwise
$$S_1(x) = \{A : A \supset C \cap [x, x+\delta) \text{ for some } \delta > 0\}.$$
Similarly define S_2 as a left version of this system. It is straightforward to verify that the pair (S_1, S_2^*) will not satisfy the joint intersection condition and yet every function has only a countable set of asymmetry points relative to this pair.

(26.7) **Example.** Using the notation of example (21.8), we may apply the above theorem to obtain an asymmetry theorem for the negligent cluster sets. For any function f the set of points
$$\{x : \Lambda N^+(f, x) \neq \Lambda N^-(f, x)\}$$
is denumerable (cf. Császár [52, Theorem 3.5, p.142]).

§27. Ambiguity theorems.

The asymmetry theorems concern the nature of the set of points x at which the two clusters set of a function at x (taken relative to two different systems S_1 and S_2) are not precisely the same. By an <u>ambiguity theorem</u> we mean an assertion about the nature of the set of points x at which, not only do the two cluster sets differ, but the two cluster sets do not have even a single cluster set value in common.

We obtain a theorem of a general type, very similar to lemma (25.3). This is closely related to a general theorem of Zajíček [268, Theorem 8, p.209].

(27.1) **LEMMA.** Let S_1 and S_2 be local systems, both of which have filtering duals. Then for any function f the amiguity set,
$$\{x : (S_1) - \Lambda(f, x) \cap (S_2) - \Lambda(f, x) = \emptyset\},$$
may be written in the form

$$\sum_{n=1}^{\infty} (S_1^* \setminus S_2)\text{-der}[A_n] \cap (S_2^* \setminus S_1)\text{-der}[B_n]$$

for some sequences of sets $\{A_n\}$ and $\{B_n\}$.

PROOF. Note firstly that if U and V are disjoint open sets such that
$$U \supset (S_1) - \Lambda(fm, x) \quad \text{and} \quad V \supset (S_2) - \Lambda(f, x)$$
then, making use of (22.4). we must have
$$f^{-1}(U) \cup \{x\} \in S_1^*(x), \quad f^{-1}(V) \cup \{x\} \in S_2^*(x)$$
$$f^{-1}(U) \cup \{x\} \notin S_2(x), \text{ and } f^{-1}(U) \cup \{x\} \notin S_1(x).$$
We may take a sequence (U_n, V_n) such that U_n, and V_n are open sets with disjoint closures and such that any pair of disjoint closed sets may be separated by some such pair. We write
$$A_n = f^{-1}(U_n) \quad \text{and} \quad B_n = f^{-1}(V_n).$$

It remains to verify that the representation in the theorem works. Let x be a point at which the two cluster sets
$$(S_1) - \Lambda(f, x) \quad \text{and} \quad (S_2) - \Lambda(f, x)$$
are disjoint. Then there must be a pair (U_m, V_m) from the sequence so that
$$U_m \supset (S_1) - \Lambda(f, x) \quad \text{and} \quad V_m \supset (S_2) - \Lambda(f, x).$$
This must then, in view of the preliminary remarks in the proof, require that x is both of the sets
$$(S_1^* \setminus S_2) \text{ der}[A_m] \quad \text{and} \quad (S_2^* \setminus S_1) \text{ der}[B_m].$$

Conversely if x is in both of these sets then, by lemma (22.4), the cluster sets must be disjoint. This completes the proof.

As an application we obtain an abstract version of a theorem of Zajíček [268]. Note that here we use the same version of a joint intersection condition as used in theorem (26.2), but stated for the duals.

(27.2) THEOREM. Let S_1 and S_2 be local systems such that both of the duals S_1^* and S_2^* are filtering and such that the pair (S_1^*, S_2^*) has the joint intersection condition of Definition (26.1). Then, for an arbitrary function f, the set
$$\{x : (S_1) - \Lambda(f, x) \cap (S_2) - \Lambda(f, x) = \emptyset\}$$
is denumerable.

PROOF. The proof is obtained as an application of the lemma. It is enough to show that for any set A, under the hypotheses here, the set of points
$$X = (S_1^* \setminus S_2) - \text{der}[A]$$
is denumerable. For each $x \in X$ define the sets

$$S_x = A \cup \{x\} \quad \text{and} \quad T_x = (\mathbb{R} \setminus A) \cup \{x\}.$$

We must have that $S_x \in S_1^*(x)$ and $T_x \in S_2^*(x)$ for each $x \in X$. But this violates the joint intersection condition if X is not denumerable and so the theorem is proved.

(27.3) **Example.** For refence we quote the theorem of Zajíček [268] which follows directly from the general theorem given above. The fact that these density systems have the appropriate intersection property is an elementary exercise. For an arbitrary function f the set of points

$$\{ x : \Lambda^+_{ess}(f, x) \cap \Lambda^-_{ess}(f, x) = \emptyset \}$$

where the two right and left essential cluster sets have no element in common must be denumerable.

§28. Essential asymmetry.

Perhaps the most interesting and most intricate of the relations that must hold between various cluster sets are obtained in the density setting. In this section we collect the known facts about such relations. Some of these can be found in the literature under the term "essential asymmetry" meant to indicate a study of the relation between the right and left essential cluster sets. Here we collect some of the information on the set of essential asymmetry (left-right) for an arbitrary function. Further information on other density cluster sets and their relations may be found in the next section.

The first attempts in this direction were to extend the known features of the ordinary right and left cluster sets to the more general density setting. The surprising fact that the set of points of left-right essential asymmetry need not be denumerable, in contrast to Young's Rome theorem, was obtained by Belowska [15] in 1960. Her proof was considered somewhat complicated and since then other simpler examples have appeared. Independently Goffman [101] and Lipiński [150] based a construction on certain properties of the Cantor set giving a relatively elementary example of a function f whose essential asymmetry points (right and left) has cardinality c. See also the accounts of Świątkowski [226] and Hunter [124]. We repeat this construction here (following the arithmetic in Świątkowski).

(28.1) **Example.** [Goffman-Lipiński] There is a function f such that the set of points of left-right essential asymmetry,

$$\{ t : \Lambda^+_{ess}(f, x) \neq \Lambda^-_{ess}(f, x) \}$$

has cardinality c.

Construction: Let $C \subset [0, 1]$ be the usual Cantor set and define A to be the union of all intervals of the form $(c-9^{-n}, c)$ such that $(c-3^{-n}, c)$ is an interval complementary to the

Cantor set. At any point x of the Cantor set the density of A at x on the right-hand side may be seen to be zero. However there are many points of the Cantor set at which the left density of A is not zero.

In fact let $\{n_k\}$ be an arbitrary increasing sequence of positive integers and define the points $^*x$ and x_j relative to this sequence:

$$x = 2 \sum_{k=1}^{\infty} 3^{-2^{n_k}},$$

and

$$x_j = -9^{-2^{n_j}} + 2 \sum_{k=1}^{j} 3^{-2^{n_k}}.$$

One checks that $x \in C$, $x_j < x$, and $x_j \to x$, and obtains the computation:

$$\frac{|A \cap (x_j, x)|}{x - x_j} \geq \frac{9^{-2^{n_j}}}{9^{-2^{n_j}} + 2 \sum_{k=j+1}^{\infty} 3^{-2^{n_k}}} \geq \frac{4}{13}.$$

Consequently at any such point x the left-hand upper density of A at x is positive. Since each sequence $\{n_k\}$ gives rise to such a point $x \in C$ there are continuum many points x in C at which the right-hand density of A is zero and the left-hand (upper) density is positive. (The cardinality of this set might also be obtained from a category argument since the set of such points x is evidently residual in C.) Taking f as the characteristic function of A, we obtain the required example.

Thus the set of points of essential (left-right) asymmetry may be large (have cardinality c). Even so it is not difficult to see that it must have measure zero. Indeed Kulbacka [139] shows that this set is also of the first category; see Goffman [101] for an easier proof. A more delicate analysis by Zajíček [268] has shown that this asymmetry set is in fact a σ-porous $F_{\sigma\delta\sigma}$ set. We give a proof here of Zajíček's theorem.

(28.2) THEOREM. [Zajíček] For an arbitrary function f the set of points of essential asymmetry

$$\{t : \Lambda^+_{ess}(f, x) \neq \Lambda^-_{ess}(f, x)\}$$

is a σ-porous Borel set of type $F_{\sigma\delta\sigma}$.

PROOF. Because of lemma (25.4) it is enough, in order to demonstrate that this set is σ-porous, to show that any set of points of the form

$$X = (S^+_{ess} \triangle S^-_{ess})\text{-der}[A]$$

is σ-porous, where we use the notation S^+_{ess} (resp. S^-_{ess}) to denote the density system that uses sets of right (left) upper density positive. This, in turn, may be established by proving that, if X is the set of points x such that the set A has exterior upper density positive at x on the right and yet has zero exterior density on the left at x, then X must be σ-porous.

For this let us write X_p ($p > 0$, p rational) for the set of points in X at which the set A has right exterior upper density exceeding p. Take a guage δ defined on X_p so that at each point x in X_p one has

$$|(x - t, x) \cap A|^e < \frac{pt}{3}$$

for all $0 < t < \delta(x)$. This function δ induces a denumerable partition $\{X_{pn}\}$ of the set X_p in our standard way.

We show that X is σ-porous now by showing that each set X_{pn} is porous at each of its points. Indeed we show that the right porosity of each X_{pn} must exceed the number $p/3$. We obtain this by a contradiction. Suppose then that the right porosity of X_{pn} is smaller than $p/3$ at some point x_0 belonging to X_{pn}. Then there must exist a positive number h_0 so that

$$\lambda(X_{pn}, x_0, x_0 + t) < \frac{pt}{3}$$

for all $0 < t < h_0$. We use this to estimate the right exterior upper density of A at x_0 and thereby obtain our contradiction.

Take any number $0 < t < h_0$ and then use the porosity requirement to find a point x_1 in the set $X_{pn} \cap (x_0, x_0 + t)$ so that

$$|x_1 - (x_0 + t)| < \frac{pt}{3}.$$

Because both the points x_0 and x_1 belong to the set X_{pn} we must have, by the nature of the partition, that

$$|A \cap (x_0, x_1)|^e < \frac{p}{3}(x_1 - x_0).$$

This gives the estimate

$$|A \cap (x_0, x_0 + t)|^e \leq |A \cap (x_0, x_1)|^e + (x_0 + t - x_1)$$
$$< \frac{p}{3}(x_1 - x_0) + \frac{pt}{3} < \frac{2pt}{3}.$$

Consequently the upper right density of A at x_0 is smaller than $2p/3$ which contradicts the fact that it was required to exceed p. From this contradiction we conclude that each set X_{pn} is porous and therefore that X is itself σ-porous. From this the first part of the theorem follows.

We wish now to establish the Borel classification of this set. By the result proved in example (23.9) the set of points (S^+_{ess})-der$[A]$ for any set A must be of type $G_{\delta\sigma}$ (where S^+_{ess} has been defined above). A similar statement may be made for the left derived set (S^-_{ess})-der$[A]$. By using these facts and the representation in lemma

(25.5) for the set of points
$$\{ x : (S^+_{ess}) - \Lambda(f, x) \neq (S^-_{ess}) - \Lambda(f, x) \}$$
we then obtain the desired result.

§29. Density cluster sets.

There are a number of other theorems available for the various density cluster sets. In the preceding section we have focused just on that relation between the right and left essential cluster sets. Here we collect some further results.

Our first theorem is again due to Zajíček [268]. The relation is established between the cluster set $H(f, x)$ (defined in example (21.4)(ii)) that uses exterior density 1, and the cluster set $M(f, x)$ (defined in example (21.5)) that uses positive outer measure.

(29.1) THEOREM. [Zajíček] Let f be an arbitrary function. Then the set of points where the exterior, upper density 1 cluster set $H(f, x)$ and the exterior positive measure cluster set $M(f, x)$ are different
$$\{ x : H(f, x) \neq M(f, x) \}$$
is of the first category.

PROOF. For the purposes of this proof let S^1 denote the system defined of sets that have upper exterior density 1 at the point, and let MS_0 denote the system that is defined so that neighbourhoods of the point have positive exterior measure. Note firstly that because of the order $S^1 \ll MS_0$ the relation
$$H(f, x) = (S^1) - \Lambda(f, x) \subset (\wedge S_0) - \Lambda(f, x) = M(f, x)$$
must hold at every point x. Thus we need to prove that the set of points
$$\{ x : (S^1) - \Lambda(f, x) \not\supseteq (MS_0) - \Lambda(f, x) \}$$
where this inclusion fails is a set of the first category. To do this we apply lemma (25.4) that exhibits this set as a union of sets of a particular form, and we see that the proof of the theorem is reduced to the following problem: show that for any set of points A the set
$$(MS_0 \setminus S^1) - \text{der} [A]$$
is first category.

By definition this is the set X of all points x such that A has positive measure in every neighbourhood of x but has upper exterior density less than 1 at x. For $0 < p < 1$ define the sets
$$X_p = \{ x \in X : A \text{ has upper ext. density at } x \text{ less than } p \}$$
so that X is a union of a denumerable number of sets X_p (for p rational and $0 < p < 1$). At each point x in X_p there is a positive number $\delta(x)$ so that
$$|A \cap (x - t, x + t)|^e < p|(x - t, x + t)|$$
for every $0 < t < \delta(x)$.

Let X_{pn} be the partition of X_p induced by the guage δ. We show that each set X_{pn} is nowhere dense. Suppose not; then there is an interval (c,d) in which X_{pn} is dense. For any points x and y $(x<y)$ in this interval (c,d) and belonging to the set X_{pn} we have the inequality
$$|A \cap (x,y)|^e < p(y-x).$$
Since X_{pn} is dense in this interval this inequality must be valid for any pair of points x and y in (c,d). In particular A cannot have any point of density 1 and so $|A \cap (c,d)| = 0$. This contradicts the fact that A must have positive measure about each point in X.

Thus we have proved that each set X_{pn} is nowhere dense and consequently that the set X, which is a denumerable union of such sets, is first category. This completes the proof.

From this theorem we obtain directly the following theorem relating the cluster sets of example (21.4)(i) and (ii).

(29.2) COROLLARY. For an arbitrary function f the set of points
$$\{ x : H(f,x) \neq \Lambda_{ess}(f,x) \}$$
is of the first category.

PROOF. This follows from the theorem and the order relation
$$S^1 \ll S^0 \ll MS_0$$
where S^1 consists of sets having exterior upper density 1, S^0 of sets having exterior upper density positive, and MS_0 of sets having positive measure in every neighbourhood.

A theorem that is similar in nature to the above results was given by Blumberg [24]. Recall that $HB(f,x)$ denotes the cluster set that uses the system of sets having (lower) exterior density 1 while $\Lambda_{ess}(f,x)$ uses the system of sets having upper exterior density positive.

(29.3) THEOREM. [Blumberg] Let f be an arbitrary function. Then the set of points
$$\{ x : HB(f,x) \neq \Lambda_{ess}(f,x) \}$$
is of measure zero.

PROOF. As in the previous theorem the proof may be reduced to the problem of showing that, for any set of points A, the set
$$(S^0 \setminus S^1) - der [A]$$
is measure zero where, for the moment $S^0(x)$, denotes the family of sets having positive upper exterior density at x, and $S^1(x)$ denotes the family of sets having exterior density 1 at x. This, in turn, reduces to showing that the set X of points x at which the set A

has positive upper density, but not density 1, is of measure zero. But this is a direct consequence of the Lebesgue density theorem and so the theorem follows. (In fact given any two sets A and X the exterior density of A is either 1 or 0 at almost every point of X; see Blumberg [24, p.265]).

(29.4)   Example.   In connection with this last example one construct a measurable function f so that the set
$$\{ x : HB(f,x) \neq \Lambda_{ess}(f,x) \},$$
which must have measure zero, is nonetheless residual. In fact, given any residual set E of measure zero, we may take f as the characteristic function of a set C that is measurable and for which the density does not exist at any point of E. (See Goffman [99].)

CHAPTER THREE

CONTINUITY

§30. Introduction.

The elementary notion of continuity has been subjected to numerous generalizations. To say that a function f is continuous at a point x is to say that, for every positive number ε, the set of points
$$\{ y : |f(x) - f(y)| < \varepsilon \}$$
is a neighbourhood of the point x. This generalizes easily; one just demands less of this set. The most useful generalizations of continuity on the real line, for the purposes of analysis, have been those of one-sided continuity and of approximate continuity, but much more delicate notions can be introduced. It is our purpose in this chapter to investigate the notion of continuity relative to a local system S and to obtain some general consequences of this concept.

Most familiar generalizations of continuity use topological ideas. The concept used here permits much weaker variants to be studied. For further general ideas on this topic see the survey article of Koszela, Świątkowski, and Wilczyński [138].

§31. S-continuity.

Relative to any local system S we define a notion of continuity that lies between ordinary continuity and feeble continuity. A function f is said to be <u>feebly continuous</u> at a point x if there is at least one sequence of points $\{x_n\}$ with $x_n \to x$ ($x_n \neq x$) and $f(x_n) \to f(x)$. Intermediate to these notions is a hierarchy of continuity notions.

(31.1) DEFINITION. Let S be a local system on a set X and let f be an arbitrary function. We say that f is <u>S-continuous</u> at the point x provided that for every $\varepsilon > 0$ the set
$$\{ t : |f(x) - f(t)| < \varepsilon \}$$
belongs to the family $S(x)$.

Note that the statement that a function f is S-continuous at a point x may also be expressed in cluster set language by the assertion that
$$f(x) \in (S)\text{-}\Lambda(f, x).$$
There are a number of elementary observations that can be made. Every continuous function is S-continuous for any system S; any S-continuous function is feebly continuous. If S_1 and S_2 are ordered so that $S_1(x) \subset S_2(x)$ then any function that is S_1-con-

tinuous at x must also be S_2-continuous there. We express this as a lemma.

(31.2) LEMMA. Let S_1 and S_2 be local systems such that $S_1 \ll S_2$. Then if a function f is S_1-continuous at a point it is also S_2-continuous there.

We will require also a generalization of the notion of semicontinuity.

(31.3) DEFINITION. Let S be a local system and let f be a function. We say that f is S-<u>upper-semicontinuous</u> at the point x if, for every $\epsilon > 0$, the set
$$\{ t : f(t) < f(x) + \epsilon \}$$
belongs to the family $S(x)$; similarly f is S-<u>lower-semicontinuous</u> at x if, for every $\epsilon > 0$, the set
$$\{ t : f(t) > f(x) - \epsilon \}$$
belongs to $S(x)$.

If a function is S-continuous then it is both S-upper-semicontinuous and S-lower-semicontinuous; if S is filtering then the converse relation is also true. Without the filtering assumption, however, a function may be semicontinuous in some sense, both above and below, and yet not be continuous in that sense. We give an elementary example to illustrate. Further problems of this type may be found in Agronsky [1].

(31.4) Example. In order that a function f be S_∞-upper-semicontinuous at a point x there must be at least one sequence $\{x_n\}$ converging to x_n ($x_n \neq x$) so that
$$\limsup_{n \to +\infty} f(x_n) \leq f(x).$$
Similarly f is S_∞-lower-semicontinuous at x if there is such a sequence $\{z_n\}$ with
$$\liminf_{n \to +\infty} f(z_n) \geq f(x).$$
Clearly a function f may be both S_∞-upper and S_∞-lower semicontinuous at a point x and not S_∞-continuous (i.e. feebly continuous) at x. For instance let f vanish at zero, be +1 at the irrationals and -1 otherwise.

There are a number of definitions in the literature of properties that might be possessed by a real function and which may be translated directly into our notion of continuity or semicontinuity. For a function f one defines the sets (called occasionally "associated sets" or "level sets" by various authors):
$$E_a = \{ y : f(y) > a \}, \quad E^b = \{ y : f(y) < b \}$$
and
$$E_a^b = \{ y : a < f(y) < b \}.$$
A function f is considered to have some property depending on whether these sets (either one of these or all of these) belong, for every pair of real numbers a and b, to some

specified family.

The assertion that a function f is S-continuous everywhere may be expressed by stating that each set $E_a^b \in S(x)$ for every pair of real numbers a and b and for every $x \in E_a^b$. Similarly the assertion that f is S-upper-semicontinuous is equivalent to the requirement that E^b belongs to $S(x)$ for every real number b and for every $x \in E^b$. Note that these ideas may also be expressed in the language of (23.1).

(31.6) Example. For a specific example of the type considered above consider the classification introduced by Zahorski [265]. A function f is an $[M_2]$-function if it is Baire 1 and each set of the form E_a or E^b has positive measure in each one-sided neighbourhood of any one of its points. If we define the system S by writing $S(x)$ for the collection of sets S for which $x \in S$, and such that for all $\epsilon > 0$, both of the sets

$$S \cap (x, x+\epsilon) \quad \text{and} \quad S \cap (x-\epsilon, x)$$

have positive exterior measure. Zahorski's notion of an $[M_2]$-function means only that f is, in addition to being Baire 1, both S-upper-semicontinuous and S-lower-semicontinuous at every point.

Similarly Zahorski's condition $[M_3]$ requires that in addition to $[M_2]$ the function be upper-semicontinuous and lower-semicontinuous at each point relative to a system described using sets that have porosity zero.

(31.7) Example. Let N be a σ-ideal of subsets of $\mathbb{R}$. A number of authors have found reasons to define a type of continuity for a function f at a point x relative to this σ-ideal by requiring that, for every $\epsilon > 0$, there is a $\delta > 0$ so that the set

$$\{ t : |f(t) - f(x)| \geq \epsilon \} \cap (x-\delta, x+\delta)$$

must belong to the class N. We shall say that f is N-continuous at x when this condition holds. If we take N as the collection of all measure zero sets (or denumerable sets, first category sets, etc.) we thus have a generalization of continuity appropriate to certain investigations.

This directly corresponds to the notion of S_N-continuity using the system S_N defined in section 11 previously.

Studies of such kinds of continuity appear in Blumberg [22] for the σ-ideals of denumerable sets, first category sets, and measure zero sets, and for a general σ-ideal in Császár [52]. We shall consider such a type of continuity in greater detail in section 34 below.

(31.8) **Example.** Parallel to this notion of **N**-continuity we have a dual notion. We say that f is <u>weakly</u> **N**-<u>continuous</u> (bilaterally) at a point x provided that for every $\epsilon > 0$ the sets

$$\{t : |f(t) - f(x)| < \epsilon\} \cap (x, x + \delta)$$

and

$$\{t : |f(t) - f(x)| < \epsilon\} \cap (x - \delta, x)$$

are not in **N** for any $\delta > 0$. Note that this is closely related to continuity with respect to a system that is dual to the S_N systems.

(31.9) **Example.** Probably the most useful kind of generalized continuity has been provided by using density considerations. A function f is said to be <u>approximately continuous</u> at a point x if for every $\epsilon > 0$ the set

$$\{t : |f(t) - f(x)| \geq \epsilon\}$$

has x as a point of dispersion (i.e. this set has exterior density 0).

From our viewpoint then this is just continuity relative to the density system S_{ap}.

This concept was first studied by Denjoy and the terminology is his. Denjoy also introduced a refinement by considering a weaker system that employs instead sets having inner, lower density exceeding 1/2 and he referred to that notion of continuity as "preponderant continuity" (intending by the term that a "preponderance" of density was required rather than full density 1). Clearly a whole hierarchy of continuity notions is available by using the various density systems.

§32. S-discontinuities.

A function is **S**-discontinuous at a point x if it fails to be **S**-continuous there. The set of **S**-discontinuities obviously is included in the set of points of ordinary discontinuity. This latter set is an F_σ set as is proved in elementary analysis courses. We cannot expect such structure in general however as the following examples will show.

(32.1) **Example.** Let E_x at each point x be the path $E_x = Q \cup \{x\}$ where Q is the set of rational numbers. Let **S** be the system generated by this system of paths (see example (13.5) for this construction). Let f be the characteristic function of the rational numbers. Then f is **S**-continuous at every rational number and **S**-discontinuous at every irrational number. Here then the set of **S**-discontinuities is not an F_σ even though this system **S** has numerous desirable properties. (For example it has an abundance of intersection properties and all the paths are dense.)

(32.2) Example. Consider the full measure system S defined by requiring that sets in S(x) all have full exterior measure in some neighbourhood of the point x. If N is any set of measure zero and f is its characteristic function then the set of points of S-discontinuity is precisely N, and N need have no particular structure.

We do however have a result that can be used to describe the relative size of the set of points of discontinuity. For quite large systems S it appears that there are "few" points of S-discontinuity. This is the theorem (24.1) that we gave previously in Chapter two and which we repeat here for emphasis.

(32.3) THEOREM. Let S be an arbitrary local system and let f be an arbitrary function. Then the set of points of S-discontinuity of the function f is a σ-(S)-negligible set.

(32.4) Example. The set of points where a function f is not at least feebly continuous is denumerable. This follows from the theorem by taking $S = S_\infty$. For other examples review section 24.

§33. Baire 1 functions.

The fact that a function f is S-continuous, even at every point, does not necessarily require that f have many nice properties. Indeed a nonmeasurable function may be everywhere S-continuous for a suitable choice of system S. If, however, the system S has some properties itself then the fact of S-continuity usually requires a great deal of the function. Our main observation in this section is that any function that is S-continuous everywhere for a system S that satisfies some intersection condition must be in the first class of Baire. This has been observed for many systems; thus functions that are continuous on one side, or approximately continuous, or preponderantly continuous, or qualitatively continuous must be Baire 1. In the literature an abundance of such theorems may be found. The theorem we now prove includes many of these as special cases.

(33.1) THEOREM. Suppose that S is a local system that satisfies an intersection condition of the form $S_x \cap S_y \neq \emptyset$. Then any function f that is everywhere both S-upper-semi-continuous and S-lower-semi-continouous must be in the first class of Baire.

PROOF. To show that f is Baire 1 it is enough, by standard arguments (Preiss [195]), to show the impossibility of the following situation. There is a closed set P and two numbers $a < b$ such that both of the sets

$$\{x \in P: f(x) < a\} \text{ and } \{x \in P: f(x) > b\}$$

are dense in P. To this end let us define the following choices
$$\{S^1_x : x \in P\} \text{ and } \{S^2_x : x \in P\}$$
from the system S. For $x \in P$, let

$S^1_x = \{y : f(y) > b\}$ if $f(x) > b$,

$S^1_x = \{y : f(y) > (a+b)/2\}$ if $f(x) = b$,

$S^1_x = \{y : f(y) < (b+f(x))/2\}$ if $f(x) < b$,

$S^2_x = \{y : f(y) < a\}$ if $f(x) < a$,

$S^2_x = \{y : f(y) < (a+b)/2\}$ if $f(x) = a$,

and
$$S^2_x = \{y : f(y) > (a+f(x))/2\} \quad \text{if } f(x) > a.$$

By the hypothesis that f is S-semicontinuous both upper and lower at each point it is clear that these are indeed choices from S. Consequently by the intersection condition there are partitions
$$\{P^1_n : n = 1, 2, \ldots\} \text{ and } \{P^2_n : n = 1, 2, \ldots\}$$
of the set P induced by the choices respectively. (We will use the properties of these partitions without comment; review section 15 for the details.)

By Baire's theorem there must be a nonempty portion $P \cap (c,d)$ in which some set $P^1_n \cap P^2_m$ is dense; we use this to obtain our contradiction. We may select points x and y in $P \cap (c,d)$ so that $f(x) > b$ and $f(y) < a$. By the nature of the partition we must be able to select further points x_1 and y_1 from the set $P^1_n \cap P^2_m$ in such a way that each of the intersections
$$S^1_x \cap S^1_{x_1}, \; S^1_{x_1} \cap S^1_{y_1}, \; S^2_{x_1} \cap S^2_{y_1}, \text{ and } S^2_{y_1} \cap S^2_y$$
is nonempty. But this is impossible; by the nature of the construction we see that $f(x_1) \geq b$ and $f(x_2) \leq a$ and the corresponding sets cannot have nonempty intersection. From this contradiction the theorem now follows.

(33.2) Example. This theorem can be used to show that functions that are everywhere continuous in any of a variety of generalized senses must be Baire 1. Since one needs only the weakest of the intersection conditions this theorem is broadly applicable. Thus functions that are approximately continuous or preponderantly continuous, even on just one side must be Baire 1.

We remark that, in the statement of theorem (33.1), the function is assumed to be both upper semicontinuous and lower semicontinuous relative to the system S. With a much stronger intersection condition this could be relaxed. Thus for the strongest system S_0 that satisfies an intersection condition of the form
$$S_x \cap S_y \cap [x,y] = [x,y]$$
the same proof shows that a function f that is everywhere (bilaterally) upper semicontinuous in the ordinary sense must be Baire 1. This was, of course, known to Baire. For

weaker systems this is not true. A function that is right upper semicontinuous or is approximately upper semicontinuous need not be Baire 1.

The intersection condition used in the theorem appears close to a necessary as well as a sufficient requirement in order that S-continuity entails Baire 1 membership. One way of exhibiting this relationship is to take the following somewhat technical construction. Let $\{S_x : x \in \mathbb{R}\}$ be some choice from a system S (i.e. each set $S_x \in S(x)$). Define an equivalence relation on $\mathbb{R}$ so that $x \leftrightarrow y$ if there is some finite sequence of points $\{z_1, z_2, z_3, \ldots, z_n\}$ so that each of the intersections

$$S_x \cap S_{z_1}, \; S_{z_1} \cap S_{z_2}, \; \ldots, \; S_{z_n} \cap S_y,$$

is nonempty. Should the system S satisfy an intersection condition then this equivalence relation must partition $\mathbb{R}$ into a denumerable family of sets $\{X_n\}$ so that pairs of elements in each X_n can be so linked. On the other hand if this partition is not denumerable then one can construct a function f that assumes an arbitrary value on each member of the partition and such a function must be S-continuous. This permits the construction then of an S-continuous function that is not Baire 1.

Another way of showing the intimate connection between the intersection condition and the Baire 1 class is provided by the next theorem.

(33.3) THEOREM. Let f be a Baire 1 function that is everywhere feebly continuous. We construct the families $S_f(x)$ as the collection of all sets S that contain a set of the form

$$\{t : |f(t) - f(x)| < \delta\} \cap (x - \delta, x + \delta)$$

for some $\delta > 0$. Then S_f is a local system that satisfies an intersection condition of the form

$$S_x \cap S_y \cap [x, y] \neq \emptyset.$$

PROOF. Since f is Baire 1 there exists, by Theorem (8.1), an interval function ψ such that

$$(S_0) - \lim \psi(x) = f(x)$$

at every point x. Let $\{S_x : x \in X\}$ be any choice from S_f. We must produce a positive function δ so that whenever

$$0 < |y - x| < \min\{\delta(x), \delta(y)\}$$

the sets S_x and S_y have a point in $[x, y]$ in common.

For this choose a positive function δ_1 on X so that

$$S_x \supset \{t : |f(t) - f(x)| < \delta_1(x)\} \cap (x - \delta_1(x), x + \delta_1(x))$$

and a positive function δ_2 on X so that

$$|\psi([x, y]) - f(x)| < \delta_1(x)/3$$

for all $0 < |y - x| < \delta_2(x)$. Now we define $\delta = \min\{\delta_1, \delta_2\}$ and show that this function gives the required intersection property.

If $|y - x| < \min\{\delta(x), \delta(y)\}$ then let δ_* denote the larger of the two values $\delta_1(x)$ and

$\delta_1(y)$: we must have

$$|\psi([x,y]) - f(x)| < \delta_1(x)/3 \quad \text{and} \quad |\psi([x,y]) - f(y)| < \delta_1(y)/3$$

so that $|f(x) - f(y)| < \delta_*$. From this, since $|x - y| < \delta_*$ we see that either y belongs to S_x or else x belongs to S_y. In either case the intersection

$$S_x \cap S_y \cap [x, y]$$

cannot be empty and so the theorem is proved.

The functions that in addition to being Baire 1 are also Darboux play a particularly important role in differentiation theory. These permit a complete characterization in this setting.

(33.4) THEOREM. Let S be a local system. Then in order that every Baire 1 function f that is S-continuous be also a Darboux function it is necessary and sufficient that S be bilateral at every point.

PROOF. This is a well known fact about Darboux Baire 1 functions and the reader is referred to the monograph of Bruckner [33, pp.8-11] for details. The key is simply that a function that is S-continuous must then be feebly continuous on both sides, and for Baire 1 functions this requires f to be Darboux. The theorem is also available for functions that are S-upper and lower semicontinous at each point, provided that S is assumed to be bilateral and to satisfy an intersection condition of the form

$$S_x \cap S_y \cap [x, y] \neq \emptyset.$$

(33.5) THEOREM. Let f be a Darboux Baire 1 function, and construct the system S_f by writing at every point x, $S_f(x)$ as the collection of all sets S that contain a set of the form

$$\{t : |f(t) - f(x)| < \delta\} \cap (x - \delta, x + \delta)$$

for some $\delta > 0$. Then S_f is a bilateral local system, that has an intersection condition of the form

$$S_x \cap S_y \cap (x, y) \neq \emptyset.$$

PROOF. This may be considered a version of a theorem of Neugebauer whose proof may be found in Bruckner [33, p.104].

§34. Negligent continuity.

We have defined in section 11 the notion of a generalized limit which ignores ("neglects") sets that belong to a specified ideal. If **N** is an ideal of sets (usually taken as a σ-ideal) that contains no interval then we say that a function f is **N**-continuous at a point x_0 provided that for every $\varepsilon > 0$ there is a $\delta > 0$ so that the set
$$(x_0 - \delta, x_0 + \delta) \setminus \{ x : |f(x) - f(x_0)| < \varepsilon \}$$
is a member of the class **N**.

For convenience we give some terminology for specific instances of this concept. For the most well known of the applications we shall use the following terminology:

(i) using **N** as the σ-ideal of denumerable sets we say that a **N**-continuous function is nearly continuous;

(ii) using **N** as the σ-ideal of measure zero sets we say that a **N**-continuous function is measure-continuous or m-continuous;

(iii) using **N** as the σ-ideal of first category sets we say that a **N**-continuous function is qualitatively continuous or q-continuous.

Note that because of (14.6) this type of continuity may be phrased in terms of path approach: f is **N**-continuous at x_0 if and only if there is a set S whose complement belongs to **N** such that
$$\lim\nolimits_{y \to x_0, y \in S} f(y) = f(x_0).$$

Such a type of continuity may be readily seen to be quite strong. Firstly if a function is **N** continuous at every point then the function is everywhere continuous in the ordinary sense.

(34.1) THEOREM. Let **N** be an ideal of sets that contains no interval. If a function f is **N**-continuous at every point then f is continuous everywhere.

PROOF. Suppose contrary to the theorem that f is discontinuous at a point x_0. Then there must be a positive number ε so that the set
$$S = (x - \varepsilon, x + \varepsilon) \setminus \{ x : |f(x) - f(x_0)| \leq \varepsilon \}$$
contains a sequence of points converging to x_0 and is a member of **N**. But this would not permit f to be **N**-continuous at the points of that sequence. From this contradiction the theorem follows.

As a consequence of this theorem it may be seen that **N**-continuity holding everywhere does not hold much interest. The most natural concept would then appear to be that of **N**-continuity that is required to hold at every point with the exception of a set that belongs to **N**. For this there are a number of equivalent expressions. (In the special case where **N** is the σ-ideal of first category sets a function f that has any of these equivalent properties is said to have the Baire property or to be "continuous apart from sets of the first category"; see Kuratowski [140, pp.399-400] or, more recently, Evans [68].)

(34.2) THEOREM. Let N be a σ-ideal of sets that contains no interval. Then for a real function f the following are equivalent.
(a) f is N-continuous everywhere except at the points of a set N_0 that belongs to N.
(b) There is a set M whose complement is in N and f is continuous relative to M at each point in M.
(c) There is a real function g with
$$\{ x : f(x) \neq g(x) \} \in N$$
and g is continuous in the ordinary sense at every point except at the points of a set N_0 that belongs to N.

PROOF. The implications (c) → (b) → (a) are immediate. Let us prove that (a) → (b). Suppose that (a) is true and let $\varepsilon > 0$. For x not in the set N_0 define
$$S_x = \{ y : |f(x) - f(y)| < \tfrac{\varepsilon}{3} \}$$
and choose the interval $I_x = (x - \delta(x), x + \delta(x))$ so that
$$N_x = I_x \setminus S_x \in N.$$
The family of sets $\{ I_x : x \in \mathbb{R} \setminus N_0 \}$ forms an open cover of the set $\mathbb{R} \setminus N_0$ and so we may find a denumerable subcover, say
$$I_{x_1}, I_{x_2}, I_{x_3}, \ldots .$$
Note that this depends on ε. Define the sets
$$N(\varepsilon) = \sum_{i=1}^{\infty} N_{x_i},$$
$$M(\varepsilon) = \mathbb{R} \setminus (N_0 \cup N(\varepsilon)).$$
The set $M(\varepsilon)$ is complementary to a set in N and the oscillation of f relative to $M(\varepsilon)$ at each point of $M(\varepsilon)$ does not exceed ε. To see this let x be any point in $M(\varepsilon)$. Then there must be an interval I_{x_i} that contains the point x and if y is any other point in $M(\varepsilon)$ also in that interval there must be a point z in S_y and in S_{x_i} so that
$$|f(z) - f(x_i)| < \tfrac{\varepsilon}{3}, \quad |f(z) - f(y)| < \tfrac{\varepsilon}{3}, \text{ and } |f(x) - f(x_i)| < \tfrac{\varepsilon}{3}.$$
from this we obtain $|f(y) - f(x)| \leq \varepsilon$ which is the result claimed above.

Now define
$$M = \prod_{n=1}^{\infty} M(\tfrac{1}{n})$$
and then one checks that M is complementary to a set in N and f is continuous relative to M at each point of M. This proves (b).

Let us now prove that (c) will follow. With M as above define the function g as
$$g(x) = \limsup_{y \to x, y \in M} f(y).$$
Then f and g agree except on a set in N and it is easy to verify that g is continuous in the ordinary sense everywhere on M. This establishes (c) and the theorem is proved.

As an application we obtain the following result for functions that are almost everywhere m-continuous. This was first obtained by Goffman.

(34.3) Example. For a real function f the following are equivalent.
(a) f is m-continuous almost everywhere.
(b) There is a set M of full measure such that f is continuous relative to M at each point in M.
(c) There is a real function g with
$$|\{ x : f(x) \neq g(x) \}| = 0$$
and g is continuous almost everywhere in the ordinary sense.

(34.4) Example. That a function is N-continuous everywhere excepting at the points of a set N_0 in N need have no consequences for the function f unless N has some strong properties (as for example if N is the class of first category sets). Indeed for an arbitrary function f there is a family N for which f is so continuous. By a theorem of Blumberg [21] there is a countable dense set C depending on f so that f is continuous relative to C at each point of C. From C such a σ-ideal N may be easily constructed.

§35. **Continuity relations.** If a function is continuous in one sense it can frequently be proved to be continuous in some other sense. We have seen some instances of this in the previous section; if a function is S-continuous then, under certain hypotheses on the system S, we have that the function f must be Baire 1 and so it is continuous in the ordinary sense at many points.

In this section we collect a variety of results each asserting that continuity in some sense will require continuity in some other sense. The first result is completely elementary.

(35.1) LEMMA. Suppose that S_1 and S_2 are two systems for which $S_1(x) \subseteq S_2(x)$ at a point x. Then, if a function f is S_1-continuous at x, it must also be S_2-continuous at that point. Conversely, if it is true that any function that is S_1-continuous at x is necessarily S_2-continuous there, then $S_1(x) \subseteq S_2(x)$.

The first of our serious results concerns the relation between S-continuity and ordinary continuity. If a function is S-continuous on an interval then, under certain hypotheses, it must be in fact continuous in the ordinary sense on that interval. The property needed in expressed in the next definition. The basic idea first appears in Świątkoskwi [226] in a general study of cluster sets and then was used by Jędrzejewski [268] in an investigation of continuity relations.

(35.2) DEFINITION. We say that a system S satisfies the condition $[J_3]$ (after Jędrzejewski) at a point x provided that whenever S is a set with $x \in S$ and such that every neighbourhood of x contains some point y for which $S \cup \{y\} \in S(y)$ then necessarily $S \in S(x)$.

This property can be expressed dually also.

(35.3) LEMMA. A system S has the property $[J_3]$ at a point x if and only if the dual S^* has the following property: for any $S \in S(x)$ there is an an interval $(x-\delta, x+\delta)$ so that
$$S \cup \{y\} \in S^*(y)$$
for all $y \in (x-\delta, x+\delta)$.

PROOF. If S has the property $[J_3]$ and there is a set S in $S^*(x)$ that does not have the property expressed in the lemma, then for some y arbitrarily close to x, $S \cup \{y\}$ is not in $S^*(y)$. This means, by definition, that $(\mathbb{R} \setminus S) \cup \{y\}$ is in $S(y)$ for such points. By the property $[J_3]$ then $(\mathbb{R} \setminus S) \cup \{x\}$ is in $S(x)$ and this is impossible because the complementary set S was assumed to belong to the dual S^*. Similar arguments show the reverse of this assertion and the lemma is proved.

We can also express this condition $[J_3]$ as a kind of intersection condition: if $S_1 \in S(x)$ then there must be a positive number δ so that for every $y \in (x-\delta, x+\delta)$ and for all $S_2 \in S^*(y)$, the intersection $S_1 \cap S_2$ contains points other than x and y. We now use this property to give the result of Jedrzejewski that shows when a continuity assumption in this setting can be used to prove continuity at a point in the ordinary sense.

(35.4) THEOREM. [Jędrzejewski] Let S be a local system that satisfies the property $[J_3]$ at the point x_0. If f is a function that is S-continuous in some deleted neighbourhood
$$(x_0 - \delta, x_0) \cup (x_0, x_0 + \delta)$$
of the point x_0 and is also S^*-continuous at the point x_0 itself, then f must be continuous in the ordinary sense at x_0.

PROOF. Suppose, contrary to the theorem, that f is not continuous at x_0. Then there must be a sequence of points $\{x_n\}$ converging to x_0 so that $f(x_n) \to y_0$ where y_0 is distinct from $f(x_0)$. We consider just the case where y_0 is finite, but the arguments can be applied to the situation in which y_0 must be taken infinite.

Set $\delta = |f(x_0) - y_0|/2$ and consider the set of points
$$S = \{t : |f(t) - f(x_0)| < \delta\}.$$
Because of the fact that f is S^*-continuous at x_0, this set belongs to $S^*(x_0)$. Define the set
$$T = \{t : y_0 - \delta < f(t) < y_0 + \delta\}.$$
By the way in which the sequence $\{x_n\}$ has been chosen it must be the case that, for sufficiently large n,
$$T \cup \{x_n\} \in S(x_n).$$
As S satisfies the condition $[J_3]$ at x_0 this means that $T \cup \{x_0\} \in S(x_0)$. But the sets S and T have no element in common other than x_0 and this is impossible if one is to be in $S(x_0)$ and the other in the dual $S^*(x_0)$. This contradiction proves the theorem.

(35.5) COROLLARY. Suppose that S is a local system, that is filtering and such that the dual S^* satisfies $[J_3]$ at each point of an open interval. Then S-continuity on that interval is equivalent to ordinary continuity there.

PROOF. If S is filtering then S-continuity implies S^*-continuity. Thus if f is S-continuous on an interval the situation of the theorem applies but using the dual system.

For a further result that gives ordinary continuity arising from S-continuity we require a slightly different version of $[J_3]$ that was given also by Jędrzejewski [268]. Note the connection with the quasi-limits defined in (13.8).

(35.6) DEFINITION. A local system S is said to satisfy the condition $[J_*]$ at a point x_0 provided that whenever S is a set with $x \in S$ and such that S contains an interval inside any neighbourhood of x_0, it must be the case that $S \in S(x_0)$.

For convenience we indicate the dual property to $[J_*]$.

(35.7) LEMMA. A system S satisfies the condition $[J_*]$ at a point x_0 if and only if the dual system S^* has the property that every set $S \in S^*(x_0)$ is dense in some neighbourhood of x_0.

PROOF. If $S \in S^*(x_0)$ is not dense in some neighbourhood of x_0 then there must be a sequence of intervals $\{(a_n, b_n)\}$ with $a_n \to x_0$, $b_n \to x_0$ and such that each intersection $S \cap (a_n, b_n)$ is empty. This would mean that the set

$$\left[\sum_{n=1}^{\infty} (a_n, b_n)\right] \cup \{x_0\}$$

cannot belong to $S(x_0)$ and this violates $[J_*]$. A similar argument works in the reverse direction.

This gives us an observation of Jędrzejewski [268] that he expressed in terms of the dual property $[J_*]$. Note that this is a converse to (34.1).

(35.8) THEOREM. If S-continuity on an interval implies ordinary continuity then each set in $S(x)$ for x in that interval must be dense in a neighbourhood of x.

PROOF. If at any point x_0 there is a set $S \in S(x_0)$ that is not dense in some neighbourhood of x_0 then there is a sequence of intervals $\{(a_n, b_n)\}$ converging to x_0 and disjoint from S. Using this fact it is easy to construct a function f that is continuous in the ordinary sense at every point other than x_0 and is S-continuous at x_0 but not in fact continuous at x_0.

These considerations may be modified to allow us to consider the relation between continuity in two general senses. Let S_1 and S_2 be two systems; a natural question is to ask for conditions under which continuity in one of the senses must imply continuity in the other sense.

(35.9) THEOREM. Let S_1 and S_2 be two systems that have this property at a point x_0 : given any $S \in S_1^*(x_0)$ and any choice

$$\{ S^2 y : y \in S, y \neq x_0 \}$$

from S_2 (i.e. each set $S^2 y \in S_2(y)$) then the set

$$\{x_0\} \cup \left[\sum_{y \in S, y \neq x_0} S^2 y\right]$$

must belong to $S_2(x_0)$.

Then, if f is S_2-continuous on an interval that includes the point x_0, f must be S_1-continuous at the point x_0.

PROOF. Suppose that f is S_2-continuous everywhere and yet not S_1-continuous at a point x_0. Then, for sufficiently small $\varepsilon > 0$, the set

$$\{ t : |f(t) - f(x_0)| < \varepsilon \}$$

cannot belong to $S_1(x_0)$. This means that the complementary set

$$S = \{x_0\} \cup \{ t : |f(t) - f(x_0)| \geq \varepsilon \}$$

belongs to $S_1^*(x_0)$ for some positive number ε.

As f is S_2-continuous everywhere we may choose for each $y \in S$ a collection of sets $\{S^2{}_y : y \in S\}$ with each $S^2{}_y \in S_2(y)$ and such that for any $z \in S^2{}_y$ the inequality $|f(z) - f(y)| < \varepsilon/2$ is valid.

By the hypothesis on the systems S_1 and S_2, the set
$$B = \{t : t = x_0 \text{ or } |f(t) - f(x_0)| > \varepsilon/2\}$$
must contain the set
$$\sum_{y \in S, y \neq x_0} S^2{}_y$$
and this latter set belongs to $S_2(x_0)$. Thus B belongs to $S_2(x_0)$ which then contradicts the fact that f must be S_2-continuous at x_0. From this contradiction the theorem follows.

For a further theorem of this type, whereby continuity in certain senses entails that in another, we give a simple intersection requirement.

(35.10) THEOREM. Let S_1 and S_2 be local systems that have this property at a point x_0: if $S_1 \in S_1(x_0)$ then for some interval $(x_0 - \varepsilon, x_0 + \varepsilon)$, for all $y \in (x_0 - \varepsilon, x_0 + \varepsilon)$, and for all $S_2 \in S_2(y)$,
$$S_1 \cap S_2 \neq \emptyset.$$
Then any function that is S_2-continuous everywhere in a neighbourhood of x_0 excepting possibly at x_0, and which is S_1-continuous at x_0 must be continuous in fact in the ordinary sense at x_0.

PROOF. If f is not continuous at x_0 then there is a sequence of points $\{x_n\}$ converging to x_0 but for which $f(x_n)$ converges to a number y_0 different from $f(x_0)$ (without loss in generality we assume y_0 is finite). Set $\varepsilon = |f(x_0) - y|/2$ and write S_1 for the set
$$S_1 = \{t : |f(t) - f(x_0)| < \varepsilon\}.$$
Since f is S_1-continuous at x_0 this set must belong to $S_1(x_0)$. Define the set S_2 as
$$S_2 = \{t : y - \varepsilon < f(t) < y + \varepsilon\},$$
and note that for sufficiently large n, $S_2 \in S_2(x_n)$. But by our hypothesis the intersection $S_1 \cap S_2$ must be nonempty and this is impossible in view of the construction. The theorem is thus proved.

CHAPTER FOUR

VARIATION OF A FUNCTION

§36. Introduction.

In this chapter we explore in greater detail the notion of variation that was introduced in chapter one. For each function f and each local system S there is an outer measure defined, $(S)\text{-}\mu_f$, that carries information as to the variation of the function f with regard to the system S. This measure can be used to discover various differentiation properties of the function, and conversely various differentiation properties of the function give rise to properties of the variational measures.

The basic definitions have appeared in the first chapter but, for easy reference, are repeated here.

(36.1) DEFINITION. Let ψ be an interval function and let C be a nonempty family of intervals. By the <u>variation</u> of ψ on C we denote

$$\text{Var}(\psi, C) = \sup \sum_{i=1}^{n} |\psi(I_i)|$$

where the supremum is taken with regard to all sequences of intervals

$$\{ I_1, I_2, I_3, \ldots \}$$

contained in C and for which, if $i \neq j$, the intervals I_i and I_j do not overlap. If C is empty we write $\text{Var}(\psi, \emptyset) = 0$.

(36.2) DEFINITION. Let S be a local system of sets and let ψ be an interval function. Then the S-variation of ψ on a set X is defined by the expression

$$V(\psi, S; X) = \inf \{ \text{Var}(\psi, C) : C \text{ an S-cover of } X \}.$$

Recall that this definition generates the following concepts.

(36.3) If ψ is an interval function and S a local system then the set function $(S)\text{-}\mu_\psi$ defined by

$$(S)\text{-}\mu_\psi(X) = V(\psi, S; X)$$

is called the <u>Stieltjes measure</u> associated with ψ and S.

(36.4) If f is a real function and Δf denotes the usual interval function

$$\Delta f : I = [x, y] \to f(y) - f(x)$$

then the set function $(S)\text{-}\mu_f$ defined by

$$(S)\text{-}\mu_f(X) = V(\Delta f, S; X)$$

is called the <u>Stieltjes measure</u> associated with the function f for the system S.

(36.5) A function f is said to have <u>zero</u> S-variation on a set X if the measure (S)-μ_f vanishes on X.

(36.6) A function f is said to have <u>finite</u> S-variation on a set X if the measure (S)-μ_f is finite on X.

(36.7) A function f is said to have σ-<u>finite</u> S-variation on a set X if the measure (S)-μ_f is σ-finite on X.

(36.8) Two interval functions ψ_1 and ψ_2 are said to be <u>S-variationally equivalent</u> on a set X provided
$$V(\psi_1 - \psi_2, S; X) = 0.$$

The fundamental property, which we have aleady obtained for this notion, asserts that for any local system S and any interval function ψ the set function (S)-μ_ψ is a metric outer measure on the real line. If, moreover, the system S is given to be filtering then this outer measure must have the increasing sets property. These facts, obviously, will permit the measures here to be used to great technical advantage. We will in this chapter explore some of the implications that these ideas have in regard to the study of various properties of real functions. In particular the variational measures play an important role in the differentiation theory of real functions.

§37. Elementary properties.

There are a number of elementary computations that will be required and we collect them in this section.

(37.1) LEMMA. Let S_1 and S_2 be two local systems ordered so that $S_1 \ll S_2$. Then for any interval function ψ one has the inequality
$$(S_1)\text{-}\mu_\psi \geq (S_2)\text{-}\mu_\psi .$$

PROOF. This follows directly from the definition since, with this order, if C is an S_1-cover of a set X then it must be also an S_2-cover of that set.

(37.2) COROLLARY. Let S be an arbitrary local system and ψ an interval function. Then the inequalities
$$(S_0)\text{-}\mu_\psi \geq (S)\text{-}\mu_\psi \geq (S_\infty)\text{-}\mu_\psi$$
must hold.

PROOF. This follows from the lemma since one has invariably the order $S_0 \ll S \ll S_\infty$.

(37.3) LEMMA. Suppose that S is a local system that is filtering, and let ψ_1 and ψ_2 be interval functions that are S-variationally equivalent on a set X. Then for every set $Y \subset X$,
$$(S)\text{-}\mu_{\psi_1}(Y) = (S)\text{-}\mu_{\psi_2}(Y).$$

PROOF. Let $\varepsilon > 0$ be given and choose an S-cover C_1 of the set X in such a way that $\text{Var}(\psi_1 - \psi_2, C_1) < \varepsilon/2$ and an S-cover C_2 of the set X in such a way that
$$(S)\text{-}\mu_{\psi_1}(X) + \frac{\varepsilon}{2} \geq \text{Var}(\psi_1, C_2).$$
Because S is given to be filtering the collection $C_1 \cap C_2$ must be an S-cover of the set X. This allows us to compute
$$(S)\text{-}\mu_{\psi_2}(X) \leq \text{Var}(\psi_2, C_1 \cap C_2)$$
$$\leq \text{Var}(\psi_2 - \psi_1, C_1 \cap C_2) + \text{Var}(\psi_1, C_1 \cap C_2)$$
$$\leq (S)\text{-}\mu_{\psi_1}(X) + \varepsilon.$$
Since ε is arbitrary this gives the inequality
$$(S)\text{-}\mu_{\psi_2}(X) \leq (S)\text{-}\mu_{\psi_1}(X)$$
and symmetrical arguments then prove the reverse inequality. The lemma follows.

(37.4) LEMMA. Let ψ_1 and ψ_2 be interval functions that are S_0-variationally equivalent on a set X. Then for any local system S the measures $(S)\text{-}\mu_{\psi_1}$ and $(S)\text{-}\mu_{\psi_2}$ are identical on the set X.

PROOF. A proof similar to that given for the preceding lemma may be used. Note that it is the strongest system S_0 that is used in hypothesis of the lemma and an arbitrary system S appears in the conclusion.

(37.5) Example. For the local systems S_0 and S_∞ and the one sided versions of them the following computations may be performed showing that these Stieltjes measures μ_f for a real function f display the size of the discontinuity of the function f at a point:
$$(S_0)\text{-}\mu_f(\{x\}) = \limsup_{t \to 0+} |f(x+t) - f(x)| + \limsup_{t \to 0-} |f(x+t) - f(x)|,$$
$$(S_0^-)\text{-}\mu_f(\{x\}) = \limsup_{t \to 0-} |f(x+t) - f(x)|,$$
$$(S_0^+)\text{-}\mu_f(\{x\}) = \limsup_{t \to 0+} |f(x+t) - f(x)|,$$
$$(S_\infty)\text{-}\mu_f(\{x\}) = \liminf_{t \to 0} |f(x+t) - f(x)|,$$
$$(S_\infty^-)\text{-}\mu_f(\{x\}) = \liminf_{t \to 0-} |f(x+t) - f(x)|,$$
and
$$(S_\infty^+)\text{-}\mu_f(\{x\}) = \liminf_{t \to 0+} |f(x+t) - f(x)|.$$

§38. Variational estimates.

For an interval function ψ and a system S there is usually an intimate connection between the values $\psi(I)$ and the variation of the function $V(\psi, S)$. The connection that is most useful to us is an estimate of the form
$$\psi(I) \leq V_I(\psi, S), \qquad (*)$$
where the variation V_I denotes the restricted variation as defined in section 19, and where ψ

is a nonnegative subadditive interval function belonging to some specified class of functions. In particular note that such an inequality puts a bound on the values of the interval function should the variation be finite, and requires the vanishing of the interval function should the variation itself vanish. Whenever an estimate such as (*) is available many consequences may be obtained, in particular monotonicity theorems are easily proved. In this section we focus just on the condition (*) itself and present some instances in which such an inequality may be present.

We state our results for ψ taken as a nonnegative subadditive interval function, that is to say

$$0 \leq \psi(I) \leq \sum_{i=1}^{n} \psi(I_i)$$

whenever $\{I_1, I_2, \ldots, I_n\}$ is a partition of an interval I. Our main interest in subsequent sections will be to apply this in the particular situations

$$\psi(I) = |\Delta f(I)|,$$

or

$$\psi(I) = \Delta f^+(I) = \max\{\Delta f(I), 0\},$$

or

$$\psi(I) = \Delta f^-(I) = \max\{-\Delta f(I), 0\},$$

for the increment of a real function f. Note that the vanishing of $|\Delta f|$ is equivalent to the fact of f being constant, and that the vanishing of Δf^- (or of Δf^+) requires f to be nondecreasing (nonincreasing). Thus we shall be able to prove constancy or monotonicity theorems directly from the estimate (*).

Our first result obtains the inequality (*) for an entirely arbitrary subadditive interval function ψ in the presence of an intersection condition on the system.

(38.1) LEMMA. Let the local system S satisfy an intersection condition of the form

$$S_x \cap S_y \cap [x, y] \neq \emptyset$$

and be bilateral at each point. Then, for any nonnegative subadditive interval function ψ and any interval I,

$$\psi(I) \leq V_I(\psi, S). \tag{*}$$

PROOF. Let C be any S-cover of the interval I. Then by theorem (16.6) there is in C a partition $\{I_i\}$ of the interval I and this gives

$$\psi(I) \leq \sum_{i=1}^{n} \psi(I_i) \leq \mathrm{Var}_I(\psi, C).$$

From this, (*) evidently follows.

(38.2) Example. This lemma applies to the systems S_0 and the density system S_{ap} both of which have been proved to satisfy such an intersection condition.

For the remainder of this section we address rather more specific concerns. The above general version of (*), proved in lemma (38.1), cannot be applied to all systems, and in particular does not apply to any one-sided systems. For certain one-sided systems we obtain some specific results that require the interval function ψ to have appropriate continuity properties in relation to the system. Our first such assertion uses the systems S_0^+, and S_0^- and their duals.

The notion of continuity that we require is expressed firstly in the definition.

(38.3) DEFINITION. Let S be a local system and let ψ be an interval function that is nonnegative and subadditive. We shall say that ψ is S-<u>continuous</u> at a point x provided that for every positive number ε there is a set $S \in S(x)$ so that
$$\psi([x, y]) < \varepsilon \text{ for all } y \in S, y \neq x.$$

(38.4) LEMMA. Let ψ be a nonnegative, subadditive interval function.
(a) If ψ is S_0^--continuous at each point then $\psi(I) \leq V_I(\psi, S_\infty^+)$.
(b) If ψ is S_∞^--continuous at each point then $\psi(I) \leq V_I(\psi, S_0^+)$.

PROOF. The proof of each of these assertions requires only an elementary compactness argument. We prove (a). Let C be an arbitrary S_∞^+-cover, and let $I = [a,b]$ be any interval. Denote by A the set of points x, $a < x \leq b$, for which
$$\psi([a, x]) \leq \text{Var}_{[a, x]}(\psi, C)$$
and set $y = \sup A$. The lemma is proved by showing that $y \in A$ and that y cannot be less than b. The continuity hypothesis supplies the former fact and the covering property supplies the latter. In a similar fashion (b) may be proved.

The above lemma uses the ordinary one-sided systems and a continuity hypothesis that arises from the opposing side. There are a number of variants of this theme we now present that use weaker systems and correspondingly stronger continuity hypotheses.

(38.5) LEMMA. Let ψ be a nonnegative, subadditive interval function.
Let S_N^+ and S_N^- denote the usual systems generated by a σ-ideal of sets N.
(a) If ψ is S_N^--continuous at each point then
$$\psi(I) \leq V_I(\psi, (S_N^+)^*).$$
(b) If ψ is $(S_N^-)^*$ - continuous at each point then
$$\psi(I) \leq V_I(\psi, S_N^+).$$

PROOF. The proof for (a) is similar to that given in lemma (38.4) above. Let C be an arbitrary $(S_N^+)^*$-cover, and let $I = [a, b]$ be any interval. Let A be the set of points x, $a < x \leq b$, for which
$$\{ t \in (a, x) : \psi([a, t]) \leq \text{Var}_{[a, t]}(\psi, C) \}$$
is not a subset of N. Again set $y = \sup A$. The lemma is proved by showing that $y \in A$ and that y cannot be less than b. As before the continuity hypothesis supplies the former fact and the covering property supplies the latter. In a similar fashion (b) may be proved.

There is a parallel assertion for density type systems. Recall that S_{ap}^+ (S_{ap}^-) denote the system of sets taken so that the right (left) interior density is 1 at the point considered; the duals $(S_{ap}^+)^*$ (and $(S_{ap}^-)^*$) then are the sets having right (left) exterior density positive at that point. Our theorem is an application of a subtle density lemma due to O'Malley; in the form given by O'Malley the function that appears is required to belong to the first Baire class, and this is why this hypothesis appears in the lemma. (According to D. Preiss this condition may not be relaxed.)

(38.6) LEMMA. Let ψ be a nonnegative subadditive interval function, let $I = [a,b]$ be an interval and suppose that the function $x \to \psi([a,x])$ is in the first Baire class.
(a) If ψ is S_{ap}^--continuous at each point then
$$\psi(I) \leq V_I(\psi, (S_{ap}^+)^*).$$
(b) If ψ is $(S_{ap}^-)^*$-continuous at each point then
$$\psi(I) \leq V_I(\psi, S_{ap}^+).$$

PROOF. We prove (a). Let C be a S_{ap}^--cover. For any interval $[a,b]$ let us define the set of points
$$A = \{ z \in (a, b] : \psi([a, z]) > \text{Var}_{[a, z]}(\psi, C) \}.$$
We shall see that such a set is always empty and from that the required inequality must follow.

Note that A is a set of type F_σ; this is because of the hypothesis on the Baire class of ψ and the fact that the other function that appears in the inequality is monotonic. At any point x in A the left density must be 1. This is because of the left continuity hypothesis: if this density is not 1 then the set of points
$$B = \{ z \in (a, b) : \psi([a, z]) \leq \text{Var}_{[a, z]}(\psi, C) \}$$
has positive upper density on the left at x, which fact, taken together with the continuity hypothesis, would require that x be in B.

Now in this situation we apply the density lemma of O'Malley [175]: if A is nonempty then there is a point x in $[a, b) \setminus A$ at which the right density of A is 1. But this is impossible for, at such a point x not in A, there must be, by the nature of the cover, positive right upper density for B at x. This contradiction supplies the proof. In a similar fashion (b) may be proved.

§39. Finite variation.

The Stieltjes measures (S)-μ_f associated with a real function f carry information as to the variation of the function f in the classical sense as well as in a number of more refined senses. For the most part the interest in such a measure is in the extreme situations in which μ_f is zero or infinite, rather than in the actual value of that measure. Thus the focus in many investigations would be on sets X for which $\mu_f(X)$ is zero, or finite, or σ-finite.

In order to appreciate how these measures behave we will in this section relate the ideas to more conventional notions of the total variation of a function. Classically the total variation of a function f on an open set G is taken as
$$\text{Var}(f, G) = \sup \left\{ \sum_{i=1}^{n} |\Delta f(I_i)| \right\}$$
where the supremum is taken with regard to all sequences $\{I_1, I_2, \ldots\}$ of nonoverlapping

intervals contained entirely in G. This relates directly to our way of computing variation.

(39.1) **LEMMA.** Let (S_0)-μ_f denote the Stieltjes measure associated with a real function f relative to the strongest local system S_0. Then, for any open set G, there is the equality
$$(S_0)\text{-}\mu_f(G) = \text{Var}(f, G)$$
for the classical variation of f on G.

PROOF. The collection C of all intervals $I \subset G$ is an S_0-cover of G and consequently
$$(S_0)\text{-}\mu_f(G) \leq \text{Var}(\Delta f, C) \leq \text{Var}(f, G).$$
Conversely if $I_1, I_2, I_3, \ldots$ is any sequence of nonoverlapping intervals contained in G and C is an S-cover of G then there must be in C a partition of each of the intervals I_i. This shows that
$$\sum |\Delta f(I_i)| \leq \text{Var}(\Delta f, C)$$
so that $\text{Var}(f, G)$ can not exceed $\text{Var}(\Delta f, C)$ for any cover C. This establishes the required equality.

Note that the classical variation is usually defined on a closed interval [a, b] as the supremum of the sums $\Sigma |\Delta f(I_i)|$ for partitions of the interval [a, b]. This variation may be seen to be given by the expression
$$\text{Var}(f, (a, b)) + |f(a+0) - f(a)| + |f(b-0) - f(b)|.$$
In particular, for a continuous function, the variation on (a, b) and on [a, b] are identical. This observation allows us to relate the classical notions directly to the variational measure here.

The lemma just given is stated for the system S_0 but may be extended to apply to any local system that has some appropriate properties.

(39.2) **LEMMA.** Let S be an arbitrary local system and f a real function. Then one has the inequality
$$(S)\text{-}\mu_f(G) \leq \text{Var}(f, G)$$
for any open set G. If S is bilateral and satisfies an intersection condition of the form
$$S_x \cap S_y \cap [x, y] \neq \emptyset$$
then one has in fact the equality
$$(S)\text{-}\mu_f(G) = \text{Var}(f, G).$$

PROOF. The proof is identical with that given for the preceding lemma; under the hypotheses here, theorem (16.6) applies to provide the needed partitions of any interval from an S-cover of that interval.

We continue with some further considerations that arise in connecting the classical ideas with the notions introduced here. If a function f has bounded variation in each interval then there may be defined a function v that describes the total variation of f in such a way that

$v(y) - v(x)$ gives the variation of f on the interval $[x,y]$. In our presentation then one might consider either of the two measures μ_f or μ_v relative to a local system S. We prove that these measures must be identical.

(39.3) THEOREM. Let f have bounded variation in each interval and let v denote its total variation function. Then the interval functions Δv and $|\Delta f|$ are S_0-variationally equivalent. In particular then for any local system S the outer measures (S)-μ_f and (S)-μ_v are identical.

PROOF. For any $\epsilon > 0$, we shall find an S_0-cover C of $\mathbb{R}$ such that
$$\text{Var}(\Delta v - |\Delta f|, C) < \epsilon.$$
To this end, for each integer $n = 0, \pm 1, \pm 2, \ldots$, we choose intervals $I_{n1}, I_{n2}, I_{n3}, \ldots I_{nm_n}$ that form a partition of the interval $[n, n+1]$ in such a way that
$$0 \leq \Delta v([n, n+1]) - \sum_{i=1}^{m_n} |\Delta f(I_{ni})| < \epsilon \, 2^{-|n|-2}.$$
Define C as the collection of all intervals which are subintervals of one or other of these intervals I_{jk}. It is clear that this collection must form an S_0-cover of the real line, and it remains only to establish the inequality posed above.

If $K_1, K_2, \ldots K_k$ is any collection of nonoverlapping intervals from this collection C then one computes, using the fact that $\Delta v(I) \geq |\Delta f(I)|$ for each interval I, that
$$\sum_{j=1}^{k} \big| \Delta v(K_j) - |\Delta f(K_j)| \big| \leq$$
$$\sum_{n=-\infty}^{+\infty} \sum_{i=1}^{m_n} \sum \left\{ \Delta v(K_j) - |\Delta f(K_j)| : K_j \subset I_{ni} \right\}$$
$$\leq \sum_{n=-\infty}^{+\infty} \sum_{i=1}^{m_n} \Delta v(I_{ni}) - |\Delta f(I_{ni})|$$
$$\leq \sum_{n=-\infty}^{+\infty} \left(\Delta v([n, n+1]) - \sum_{i=1}^{m_n} |\Delta f(I_{ni})| \right)$$
and by hypthesis this is less than ϵ. As this holds for each choice of intervals $\{K_1, K_2, \ldots\}$ from C the required inequality has been proved.

The final assertion of the theorem is then a consequence of lemma (37.4).

From this result we may readily determine how the measures behave with regard to the usual Jordan decomposition of a function of bounded variation. It is natural to ask also for the situation that holds for other classical decompositions. If a function f is nondecreasing then it may be written uniquely as $f = f_c + f_d$ where the functions f_c and f_d are nondecreasing, f_c is continuous and f_d is the saltus of f. Since outer measures permit similar decompositions one should expect that there is a close connection in our setting. In fact we have the following.

(39.4) LEMMA. Let the function f be nondecreasing and let $f = f_c + f_d$ be the decomposition of f into its continuous and discrete parts. Write D_f for the set of points of discontinuity of f. Then for any system S,

(i) $(S)\text{-}\mu_f = (S)\text{-}\mu_{f_c} + (S)\text{-}\mu_{f_d}$,

(ii) $(S)\text{-}\mu_f = (S)\text{-}\mu_{f_c}$ on the set $\mathbb{R} \setminus D_f$, and

(iii) for any set X,

$$(S)\text{-}\mu_{f_d}(X) = \sum_{x \in X \cap D_f} (S)\text{-}\mu_f(\{x\}).$$

PROOF. To simplify the writing set $g = f_c$ and $s = f_d$. Then $f = g - s$ so that we may compute
$$V(\Delta f - \Delta g, S_0, \mathbb{R} \setminus D_f) = V(\Delta s, S_0, \mathbb{R} \setminus D_f).$$
Simple arguments allow us to compute for this discrete function s that
$$V(\Delta s, S_0, \mathbb{R} \setminus D_f) = 0$$
and so
$$V(\Delta f - \Delta g, S_0, \mathbb{R} \setminus D_f) = 0.$$
From lemma (37.4) then we must have for any system S that the measures $(S)\text{-}\mu_f$ and $(S)\text{-}\mu_g$ must agree on the set $\mathbb{R} \setminus D_f$ as required.

Similarly for any point x in D_f, using the continuity of g
$$V(\Delta f - \Delta s, S_0, \{x\}) = V(\Delta g, S_0, \{x\}) = 0.$$
From this we see that assertion (iii) of the lemma must be true. These two together give (i) and the lemma is proved.

A similar type of theorem is available for the Lebesgue decomposition of a function of bounded variation. If the nondecreasing function f is written as the sum $f = f_{ac} + f_{sing}$ for a function that is absolutely continuous and a function that is singular then again a decomposition of the measures relative to a system S is induced. This material will be developed and extended in sections 45 and 47.

§40. σ-finite variation.

The results given in the preceding section apply to functions that have finite variation in the classical sense. For more delicate variational ideas we need to turn to the notion of a function that is of generalized bounded variation (VBG*). The definitions may be found in Saks [209,p.228] or Bruckner [33,pp.82-83]. The main result provides the connection between these two concepts. We know that a function has bounded variation on an interval if and only if the variational measure is finite there. As we shall see, it is very nearly the case that a function is VBG* on a set if and only if the variational measure $(S_0)\text{-}\mu_f$ is σ-finite on that set. Notice, however, that if a function f is VBG* on a set X then it is also VBG* on the set $X \cup N$ for any denumerable set N. On the other hand our measure μ_f cannot be σ-finite unless it assigns finite measure to any singleton set; thus the property of the measure being σ-finite is, unlike the VBG* property, sensitive to the addition of denumerable sets.

(40.1) THEOREM. Let $(S_o)\text{-}\mu_f = \mu_f$ denote the Stieltjes measure associated with a real function f relative to the strongest system S_o. Then μ_f is σ-finite on a set X if and only if f is VBG* on that set f and
$$\limsup_{t \to 0} |f(x+t)| < +\infty$$
at each point x in X.

PROOF. The requirement that f have finite extreme limits at each point of X in order that μ_f be σ-finite on X is clearly necessary, in view of the computations given in example (37.5) above. To prove the lemma in one direction, let us show that, whenever $\mu_f(X) < +\infty$, it must follow that f is VBG* on X. This will prove that the σ-finiteness of f on a set entails that f is VBG* there.

Suppose that an S_o-cover of the set X has been found for which
$$\text{Var}(\Delta f, C) < +\infty .$$
There must then be a choice of sets $\{S_x : x \in X\}$ from S_o so that for every $y \in S_x$ ($y \neq x$) the interval [x, y] belongs to C. We use an intersection condition on S_o here to induce a partition $\{X_n\}$ of the set X in such a way that for x and y (x < y) in the same member X_n one must have
$$S_x \cap S_y \cap [x, y] = [x, y] .$$
Now, if $\{[x_i, y_i]\}$ is any sequence of nonoverlapping intervals with endpoints in a set X_n, we may check, for the sum of the oscillations, that
$$\sum O(f, [x_i, y_i]) \leq 2 \text{Var}(\Delta f, C) < +\infty .$$
Thus f is VB* on each set X_n, and consequently f is VBG* on X as required.

Conversely suppose that f is VB* on a set X so that (using the notation of Saks [209, p.228])
$$V_*(f, X) < +\infty ,$$
and suppose that
$$\limsup_{t \to 0} |f(x+t)| < +\infty$$
at each point of X. (Since f is VB* on X there would be only denumerably many points at which this might be infinite.) Let Y be the set of points in X that are isolated on one side at least in X, and take C as the collection of all intervals that have one endpoint in $X \setminus Y$ and lie within the bounds of X. Then we compute
$$\mu_f(X \setminus Y) \leq \text{Var}(\Delta f, C) \leq 2 V_*(f, X) < +\infty .$$
Since $\mu_f(\{x\}) < +\infty$ at each point x in Y we see that μ_f must be σ-finite on X as required. If f is merely VBG* on X the same conclusion evidently holds and so the lemma is proved.

We see from this theorem that there is an intimate relation between the variational measure $(S_o)\text{-}\mu_f$ and the classical notions of variation for the function f. This measure carries the information as to where the function f has bounded variation and as to where it has generalized bounded variation (VBG*). For other systems S the corresponding measure will offer a more general and delicate estimate of the variation of the function. In a great many cases there is a connection available with the notion VBG which is much weaker than the VBG* idea. This is given in the next theorem, and requires only an appropriate intersection condition. The definition of VBG is given in Saks [209,p.221]; "square-bracket" VBG (i.e. [VBG]) indicates that in the sequence of sets on each of which the given function is to be VB, the sets may be taken to be closed. The assertion that a function f is [VBG] on a set X means, then, that f is VB on each of a sequence of sets X_n whose union is X and with X_n closed relative to the set X. If X is closed each set X_n is simply closed in the usual sense.

(40.2) THEOREM. Let S be a local system that has an intersection condition of the form
$$S_x \cap S_y \cap [x, y] \neq \emptyset$$
on a set X. Then if the measure $(S)-\mu_f$ is σ-finite on the set X it follows that f must be [VBG] on X.

PROOF. We may suppose that $(S)-\mu_f(X) < +\infty$ and show then that f is VBG on the set X. (The refinement whereby f is [VBG] on X follows after.) We take an S-cover C of the set X so that
$$\text{Var}(\Delta f, C) < +\infty$$
and then a choice $\{S_x : x \in X\}$ so that C contains all the intervals [x,y] for which $y \in S_x$.

By the intersection condition there must be a partition $\{X_n\}$ of the set X so that whenever x and y belong to the same member X_n of the partition the intersection
$$S_x \cap S_y \cap [x, y]$$
is nonempty. Suppose that $[x_1, y_1], [x_2, y_2], [x_3, y_3], \ldots$ is a sequence of nonoverlapping intervals with endpoints in X_n. Then we may use the intersection condition to determine points $z_1, z_2, z_3, \ldots$ in the appropriate intersections so that

$$\sum |f(y_i) - f(x_i)| \leq \sum |f(y_i) - f(z_i)| + |f(z_i) - f(x_i)|$$
$$\leq \text{Var}(\Delta f, C) < +\infty .$$

This inequality shows that f is VB on each set X_n and consequently f must be VBG on X.

We will now check that f is in fact VB on the closure of each set X_n in X which will then verify that f is [VBG] on X as required. Suppose that $[x_1, y_1], [x_2, y_2], [x_3, y_3], \ldots$ is a sequence of nonoverlapping intervals with endpoints in the closure of X_n taken in X. Then we may find points $a_1, b_1, a_2, b_2, \ldots$ in X_n sufficiently close to the points $x_1, y_1, x_2, y_2, \ldots$ respectively so that the intersections
$$S_{x_i} \cap S_{a_i} \cap [x_i, a_i]$$
and
$$S_{y_i} \cap S_{b_i} \cap [y_i, b_i]$$
are nonempty. These points may be chosen in such a way as to permit a repetition of the argument above to obtain that

$$\sum |f(y_i) - f(x_i)| \leq 3 \text{ Var}(\Delta f, C) .$$

Thus f is VB on $\overline{X}_n \cap X$ as required to complete the proof.

§41. The Vitali theorem.

The measures $(S)\text{-}\mu_f$, that we are using to describe the total variation of a function, may be considered to be generalizations of the classical Stieltjes measures associated with any function of bounded variation. In this section we show the precise connection. For discontinuous functions f our measures treat the discontinuities in a different way than is usual and so the theorem is best viewed just for continuous f. Since any monotonic function f may be written as a sum $f = f_c + f_d$ of a continuous monotonic function and a discrete monotonic function and since the associated Lebesgue-Stieltjes measures have a similar decomposition this focus on continuous functions will, nonetheless, allow us to draw more general conclusions.

For a continuous, nondecreasing function f the standard Lebesgue-Stieltjes outer measure λ_f may be simply defined, for bounded open sets G, as

$$\lambda_f(G) = \sum \Delta f(I_i)$$

where the sum is taken over all of the component intervals $\{I_n\}$ of G, and defined for arbitrary bounded sets X as

$$\lambda_f(X) = \inf\{\lambda_f(G) : G \text{ open}, G \supset X\}.$$

Our main theorem asserts that this outer measure λ_f is identical to the variational measure $(S)\text{-}\mu_f$ for any local system S and for any continuous nondecreasing function f. This is simply a consequence of the Vitali theorem for Lebesgue-Stieltjes measures and we shall so label it.

(41.1) THEOREM. [Vitali] Let f be a continuous and nondecreasing function and let λ_f denote its associated Lebesgue-Stieltjes measure. Then, for any local system S, the two measures λ_f and $(S)\text{-}\mu_f$ are identical.

PROOF. We restrict the argument to bounded sets from which the general result evidently follows. If G is a bounded open set and $\{(a_k, b_k)\}$ are the component intervals of G then we know that

$$(S_0)\text{-}\mu_f(G) = \sum f(b_k) - f(a_k) = \lambda_f(G).$$

Thus for any bounded set X, and G open, $G \supset X$,

$$(S_0)\text{-}\mu_f(X) \leq (S_0)\text{-}\mu_f(G) = \lambda_f(G).$$

Hence, by the definition of λ_f, we have

$$(S_0)\text{-}\mu_f \leq \lambda_f.$$

On the other hand let C be an S_∞-cover of a bounded set X (that is to say then that C is a Vitali cover of X) and apply the Vitali theorem for Lebesgue-Stieltjes measures λ_f to obtain a sequence $\{I_1, I_2, \ldots\}$ of nonoverlapping intervals from C so that

$$\lambda_f\left(X \setminus \sum_{i=1}^{\infty} I_i\right) = 0.$$

In particular we have

$$\lambda_f(X) \leq \sum_{i=1}^{\infty} \lambda_f(I_i) = \sum_{i=1}^{\infty} \Delta f(I_i) \leq \mathrm{Var}(\Delta f, C).$$

As this holds for all such C this gives the inequality

$$\lambda_f \leq (S_\infty)\text{-}\mu_f.$$

But we have always that $(S_0)\text{-}\mu_f \geq (S_\infty)\text{-}\mu_f$ so that the equality of the three measures

$$\lambda_f, (S_0)\text{-}\mu_f, \text{ and } (S_\infty)\text{-}\mu_f$$

has been established. For any local system S the measure $(S)\text{-}\mu_f$ lies between the measures $(S_0)\text{-}\mu_f$ and $(S_\infty)\text{-}\mu_f$ and so the theorem is proved.

(41.2) COROLLARY. Let f be a continuous function that has bounded variation in any interval, let v denote its total variation function and let λ_v denote the Lebesgue-Stieltjes measure associated with v. Then for any local system S the measures

$$(S)\text{-}\mu_f, (S)\text{-}\mu_v, \text{ and } \lambda_v$$

are identical.

PROOF. This follows from the theorem together with theorem (39.3) above.

These results may also be interpreted by the following assertion.

(41.3) COROLLARY. Let f be a continuous function that has bounded variation in any interval, let v denote its total variation function and let λ_v denote the Lebesgue-Stieltjes measure associated with v. Then the measures

$$(S_0)\text{-}\mu_f, (S_0)\text{-}\mu_v, (S_\infty)\text{-}\mu_f, (S_\infty)\text{-}\mu_v \text{ and } \lambda_v$$

are identical.

In particular, for the function $f(x) = x$, each of these measures is equivalent to Lebesgue outer measure. For many applications the Vitali theorem, which in our version asserts that the two measures $(S_0)\text{-}\mu_f$ and $(S_\infty)\text{-}\mu_f$ are identical, must be extended to a larger class of functions. We prove in our next theorem that a similar assertion is available for functions that are VBG∗.

(41.4) THEOREM. Let the function f be VBG∗ on a set E and let C_f denote the set of points at which f is continuous. Then the measures $(S_0)\text{-}\mu_f$ and $(S_\infty)\text{-}\mu_f$ are identical on the set $E \cap C_f$.

PROOF. To begin let us suppose that f is VB∗ on a bounded closed set E and write a, b for the bounds of E and $\{I_k\}$ for the sequence of intervals contiguous to E in [a,b]. We know that the sequence of oscillations

$$\sum_{k=1}^{\infty} O(f; I_k) < +\infty.$$

must converge. Define the function g so that $f(x) = g(x)$ for every x in E, so that

$g(x) = f(a)$ for $x < a$, so that $g(x) = f(b)$ for $x > b$, and finally so that g is linear on the contiguous intervals $\{I_k\}$. Note that the function g has bounded variation and is continuous at each point in E at which f is continuous. We shall show that
$$V(\Delta f - \Delta g, S_0, E \cap C_f) = 0.$$
Let A denote the set of points in E that are not isolated on either side. Since $E \setminus A$ is denumerable and the functions f and g are continuous on C_f we see that
$$V(\Delta f - \Delta g, S_0, (E \setminus A) \cap C_f) = 0$$
and so we may concentrate on the subset A.

Let $\epsilon > 0$ and choose an integer N so large that
$$\sum_{k=N+1}^{\infty} O(f; I_k) < \frac{\epsilon}{2}.$$
Now select an S_0-cover C of the set A so that every interval $I \in C$ fails to overlap an interval of the sequence I_k for $k \leq N$. Now we claim that
$$\text{Var}(\Delta f - \Delta g, C) \leq \sum_{k=N+1}^{\infty} 2 O(f; I_k) < \epsilon.$$
To see this note that if $I = [c,d]$ is an interval in C with endpoints in the set E then
$$\Delta f(I) - \Delta g(I) = 0.$$
If however $c \in E$ but d belongs to a contiguous interval I_k then
$$|\Delta f(I) - \Delta g(I)| \leq 2 O(f; I_k)$$
and $k \geq N+1$. A similar argument applies if d is in E and c is not. Thus
$$V(\Delta f - \Delta g, S_0, A) = 0.$$
as required.

From the fact that
$$V(\Delta f - \Delta g, S_0, E \cap C_f) = 0$$
we may conclude that the measures $S_0\text{-}\mu_f$ and $S_0\text{-}\mu_g$, and the measures $S_\infty\text{-}\mu_f$ and $S_\infty\text{-}\mu_g$ agree on the set $E \cap C_f$. By (41.2) the measures $S_0\text{-}\mu_g$ and $S_\infty\text{-}\mu_g$ are identical on the set C_f. Consequently the theorem is proved in this special case.

In general if f is VBG$_*$ on a set E then there is a sequence of closed bounded sets $\{E_n\}$ whose union includes E and on each set E_n f is VB$_*$. Now the two measures of the theorem agree on each subset of $E_n \cap C_f$ and elementary properties of the measures show that they must agree on each subset of $E \cap C_f$.

§42. Derivates and variation.

There are many connections between the variation of a function and the size of its derivates. Thus, for example, a function with bounded derivates may be shown to be Lipschitz and so to have bounded variation; in the other direction a function of bounded variation has finite derivates almost everywhere. These elementary results have been generalized in numerous ways. We present a unified treatment of some of this material in the setting of a local system.

Given a local system S and a function f we wish to determine relations that must hold between the variation $(S)\text{-}\mu_f$ and the derivates $(S)\text{-}\underline{D} f(x)$ and $(S)\text{-}\overline{D} f(x)$. The basic computations are presented in this section.

(42.1) LEMMA. Let S be a system that is filtering and suppose that, at each point x in a set X, one has the inequalities
$$- M \leq (S)\text{-}\underline{D}\, f(x) \leq (S)\text{-}\overline{D}\, f(x) \leq M .$$
Then
$$(S)\text{-}\mu_f (X) \leq M\, |X| .$$

PROOF. For every positive number ε consider the family of intervals
$$C_1 = \{\, I : |\Delta f(I)| < (M + \varepsilon)\, |I|\, \} .$$
Clearly, because of the assumed inequalities in the statement of the lemma, this family C is an S-cover of the set X. Consequently, if C_2 is an arbitrary S-cover of the set X, we must have
$$(S)\text{-}\mu_f (X) \leq \mathrm{Var}\,(\Delta f, C_1 \cap C_2) \leq (M + \varepsilon)\, \mathrm{Var}\,(\Delta x, C_2) .$$
But $|X| = \inf \mathrm{Var}\,(\Delta x, C_2)$ where the inf is taken for all such collections C_2. Since $\varepsilon > 0$ is arbitrary, the final statement of the lemma now follows.

(42.2) LEMMA. Suppose that $M > 0$ and that everywhere on a set X one at least of the the inequalities
$$(S)\text{-}\overline{D}\, f(x) < -M \quad \text{or} \quad (S)\text{-}\underline{D}\, f(x) > M$$
holds. If S is filtering, then
$$(S)\text{-}\mu_f (X) \geq M\, |X| .$$
If S has a filtering dual, then
$$(S^*)\text{-}\mu_f (X) \geq M\, |X| .$$

PROOF. Suppose firstly that S is filtering and choose, for any positive number ε, an S-cover C_1 of X such that
$$\mathrm{Var}\,(\Delta f, C_1) \leq (S)\text{-}\mu_f (X) + \varepsilon .$$
Define the collection of intervals
$$C_2 = \{\, I : |\Delta f(I)| > M\, |I|\, \}$$
which, because of the inequalities given in the assertion of the lemma, must be an S-cover of the set X. Since S is filtering $C = C_1 \cap C_2$ must also be an S-cover of X and this gives the computations
$$(S)\text{-}\mu_f (X) + \varepsilon \geq \mathrm{Var}\,(\Delta f, C) \geq M\, \mathrm{Var}\,(\Delta x, C) \geq M\, |X|$$
and the desired inequality is proved.

If on the other hand S^* is filtering then choose instead C_1 as an S^*-cover of X and repeat the above arguments to obtain the desired inequality.

(42.3) THEOREM. Suppose that S is a local system that is filtering and that the measure $(S)\text{-}\mu_f$ is σ-finite on a set X. Then, at almost every point x of the set X, the inequalitites
$$-\infty < (S)\text{-}\underline{D}\, f(x) \leq (S)\text{-}\overline{D}\, f(x) < +\infty$$
must hold. Conversely, if these inequalities hold everywhere on the set X, then the measure $(S)\text{-}\mu_f$ must be σ-finite on that set.

PROOF. Let Y denote any set of points x at which the lower derivate $(S)\text{-}\underline{D}\,f(x)$ is equal to $-\infty$. Then at every such point x and for every positive number M,
$$(S)\text{-}\underline{D}\,f(x) = (S^*)\text{-}\overline{D}\,f(x) < -M.$$
By the estimate in lemma (42.2) this gives
$$M\,|Y| \le (S)\text{-}\mu_f(Y).$$
Since the measure $(S)\text{-}\mu_f$ is σ-finite on the set X of the theorem it will follow from the above estimate that
$$|\{\,x \in X : (S)\text{-}\underline{D}\,f(x) = -\infty\,\}| = 0$$
exactly as required. Similar arguments handle the upper derivate so that the first part of the theorem is proved.

In the other direction, if these derivates are finite, we write
$$X_n = \{\,x \in X \cap [-n,n] : -n \le (S)\text{-}\underline{D}\,f(x) \le (S)\text{-}\overline{D}\,f(x) \le n\,\}.$$
Lemma (42.1) gives
$$(S)\text{-}\mu_f(X_n) \le n\,|X_n| < +\infty$$
so that $(S)\text{-}\mu_f$ is finite on each set X_n. Since X is the union of this sequence of sets, $(S)\text{-}\mu_f$ must be σ-finite on X as required.

Parallel to this theorem we have a similar result for zero derivates and zero variation.

(42.4) THEOREM. Suppose that S is a local system that is filtering and that the measure $(S)\text{-}\mu_f$ vanishes on a set X. Then, at almost every point x of the set X, the (S)-derivative of f must vanish,
$$(S)\text{-}D\,f(x) = 0.$$
Conversely, if the (S)-derivative of f vanishes everywhere on the set X, then the measure $(S)\text{-}\mu_f$ must vanish on that set.

PROOF. The details are almost identical to those of the preceding theorem and we leave them to the reader.

§43. Further measure estimates.

In the literature the Lebesgue-Stieltjes measure λ_f is frequently used in situations in which f has bounded variation. For a well known example a function f is absolutely continuous in the classical sense if and only if f has bounded variation and the measure λ_f vanishes on each set of Lebesgue measure zero (i.e. λ_f is absolutely continuous in the measure-theoretic sense). For functions f that do not have bounded variation one is more likely to encounter the set function
$$E \to |f[E]| = |\{\,x : f(t) = x \text{ for some } t \in E\,\}|$$
and many theorems have consequences that hold with the exception of a set N for which this image f[N] has measure zero.

In our setting it is more natural to utilize the variational measures $(S)\text{-}\mu_f$. Consequently it is of some interest to relate these measures with the "image" measure above. In this section we explore some such relations.

Our first result gives the inequality
$$|f[E]| \leq (S_0)\text{-}\mu_f(E).$$
In fact this inequality is available for any system S that satisfies an elementary intersection condition.

(43.1) THEOREM. Let S be a local system that satisfies an intersection condition of the form
$$S_x \cap S_y \cap [x, y] \neq \emptyset.$$
Then for any set E and any real function f the inequality
$$|f[E]| \leq (S)\text{-}\mu_f(E)$$
must hold.

PROOF. Let C be an arbitrary S-cover of the set E. This means that there must exist a choice
$$\{ S_x : x \in E \}$$
from S such that each interval $[x,y] \in C$ whenever $y \in S_x$ $(y \neq x)$. Thus, by the intersection condition, there must be a guage δ on E so that whenever
$$0 < |y - x| < \min \{\delta(x), \delta(y)\}$$
the intersection
$$S_x \cap S_y \cap [x, y]$$
is nonempty. Define the sequence of sets $\{E_n\}$ by setting
$$E_n = \{ x \in E : \delta(x) > \frac{1}{n} \} \qquad n = 1, 2, 3, \ldots.$$
Observe that this is an expanding sequence of sets whose union is E.

Let $\{I_k\}$ be an enumeration of the intervals
$$[\frac{m}{n}, \frac{m+1}{n}) \quad m = 0, \pm 1, \pm 2, \ldots$$
where n is held fixed. For any pair of points x and y $(x < y)$ that belong to a set $E_n \cap I_k$ there must be a point z between x and y so that each of the intervals $[x,z]$ and $[z,y]$ belongs to C. From this observation we obtain the estimate
$$|f[E_n]| \leq \sum_{k=1}^{\infty} |f[E_n \cap I_k]| \leq \text{Var}(\Delta f, C).$$
Since this holds for every cover C we obtain
$$|f[E_n]| \leq (S)\text{-}\mu_f(E)$$
and then, taking the limit for the expanding sequence $\{E_n\}$, we have the inequality of the theorem as required.

(43.2) THEOREM. [Saks-Sierpinski] Let f be an arbitrary function and let ψ denote the image "measure"
$$\psi(E) = |f[E]|.$$
Then for any bounded set E_0
$$|f[E_0]| \leq 2(S_\infty)\text{-}\mu_\psi(E_0).$$

PROOF. See Saks [209, p.211].

A further estimate of this type has appeared in Thomson [233]. We have seen that the vanishing of the measure (S)-μ_f on a set X will require, under certain mild hypotheses, that the image of X under f, i.e. the set $f[X]$, has measure zero. There is a converse available to this whereby, if f is continuous, the vanishing of $|f[X]|$ will require the vanishing of the measure (S_∞)-$\mu_f(X)$.

(43.3) THEOREM. Let f be a continuous function and X a set of real numbers for which $|f[X]| = 0$. Then
$$(S_\infty) - \mu_f(X) = 0 .$$

PROOF. We may suppose that the set X is a subset of the interval $[0,1)$ as the general case follows at once from this particular case. Define now the following subsets of that interval:

$$I_{jn} = \left[\frac{j-1}{2^n}, \frac{j}{2^n} \right) \quad n = 1,2,\ldots; j = 1,2,\ldots,2^n ,$$

$$B_{jn} = \{ x \in X \cap I_{jn} : f(x) = f(x') \text{ for some } x' \in I_{jn} \},$$

$$B_n = \sum_{j=1}^{2^n} B_{jn} ,$$

$$B = \prod_{n=1}^{\infty} B_n ,$$

$$D_n = X \setminus B_n$$

and

$$D = \sum_{n=1}^{\infty} D_n .$$

By the nature of this construction we must have
$$X \subset D_n \cup B_n$$
for each index n, so that in fact
$$X \subset D \cup B .$$
Consequently to prove that the measure (S_∞)-μ_f vanishes on the set X it is enough to show that it vanishes on both of the sets D and B.

On each set $D_n \cap I_{jn}$ one checks that the function f is necessarily monotonic and its values lie in a set of measure zero. From this it is easy to show that
$$(S_\infty) - \mu_f(D_n \cap I_{jn}) = 0 .$$
This expresses D as a countable union of sets each of which has (S_∞)-μ_f measure zero,
$$D = \sum_{n=1}^{\infty} \sum_{j=1}^{\infty} D_n \cap I_{jn} ,$$
and so our measure must vanish on D also.

Finally consider the set B. By the construction the function f is locally recurrent at each point of B; that is at every point $x \in B$ there is some sequence of points x_n, different from x and converging to x, so that $f(x_n) = f(x)$. Consequently there is an S_∞ cover C of the set B such that $\Delta f(I) = 0$ for every $I \in C$. From this we see that the measure (S_∞)-μ_f vanishes on B

as well and the theorem is proved.

(43.4) COROLLARY. Let f be a continuous function that is VBG* on a set X. Then, for any subset N of X, the measure $|f[N]|$ vanishes if and only if the measure $(S_0)-\mu_f(N)$ vanishes.

PROOF. Because of theorem (43.1) whenever $(S_0)-\mu_f(N) = 0$ it must be the case that $|f[N]| = 0$. Because of theorem (43.3) whenever $|f[N]| = 0$ the measure $(S_\infty)-\mu_f(N) = 0$. But for continuous VBG* functions these two measures are identical on subsets of X and so the corollary follows.

§44. The Lebesgue-Denjoy-Lusin theorem.

Lebesgue obtained in 1904 the fundamental result that functions of bounded variation are almost everywhere differentiable. Later Denjoy and Lusin showed, independently, that this same conclusion was available for functions that are VBG**. Over the years many proofs have been given for this result of Lebesgue, some of them very elegant and simple. The proof that we present here uses the Vitali theorem in a particularly simple way, but the proof itself cannot be considered elementary since it must depend on the Vitali theorem. Note that we may pass immediately to a consideration of the more general version of the theorem wherein the generalized variation (VBG*) is assumed.

(44.1) THEOREM. Let the function f be VBG* on a set E. Then f must have a finite derivative f'(x) at almost every point of E.

PROOF. We obtain a proof by investigating the set of points X_{pqr} defined as
$$X_{pqr} = \{ x \in E : p > \overline{D} f(x) > q > r > \underline{D} f(x) \}$$
for rational numbers p,q,r with $p > q > r$ and as
$$X_{pqr} = \{ x \in E : \overline{D} f(x) > r > q > \underline{D} f(x) > p \}$$
for rational numbers p, q, r with $p < q < r$.

Note that, because f is VBG*, the measure $(S_0)-\mu_f$ is σ-finite on E (excepting a denumerable set) and so the derivates are, by theorem (42.3), finite a.e. on the set E. Consequently almost every point x at which the derivative f'(x) does not exist as a finite number must appear in one of these sets.

The theorem is proved by showing that each such set has zero measure. Without loss of generality we may ignore the denumerable collection of points of discontinuity of f and assume that f is both continuous and VBG* on E. Take the set X_{pqr} with $p < q < r$ and define the function $f_p(x) = f(x) - px$. At each point x in X_{pqr} we have
$$\overline{D} f_p(x) > r - p > 0 \quad \text{and} \quad q - p > \underline{D} f_p(x) > 0.$$
From this we obtain by an elementary argument (cf. the proof of (42.2)) that
$$(S_0)-\mu_{f_p}(X_{pqr}) \geq (r - p)(S_\infty)-\mu_x(X_{pqr})$$
and

$$(S_\infty)\text{-}\mu f_p (X_{pqr}) \leq (q - p) (S_0)\text{-}\mu_x (X_{pqr}).$$

Since f_p is continuous and VBG∗ on E the measures $(S_0)\text{-}\mu f_p$ and $(S_\infty)\text{-}\mu f_p$ agree on E, as do the two μ_x measures which we know to be precisely Lebesgue outer measure. Since $q < r$ this is possible only if X_{pqr} has measure zero. Note too that the measure $(S_0)\text{-}\mu f$ must also vanish on these sets; this fact is needed later. Since each of these sets is seen to have measure zero the exceptional set of the theorem must have measure zero and the theorem is proved.

The exceptional set where the derivative does not exist may be analysed somewhat further by using the measure $(S_0)\text{-}\mu f$.

(44.2) THEOREM. Let f be a continuous function that is VBG∗ on a set E. Then f has a derivative f'(x), either finite or infinite, $(S_0)\text{-}\mu_f$ almost everywhere in E.

PROOF. Let X be the set of points x in E at which f'(x) does not exist (including $+\infty$ or $-\infty$). This may be expressed as the the union of three sets X_1, X_2 and X_3 where

$$X_1 = \{ x \in E : f'_+ (x) = +\infty \text{ and } f'_- (x) = -\infty \}$$
$$X_2 = \{ x \in E : f'_+ (x) = -\infty \text{ and } f'_- (x) = +\infty \}$$

and X_3 is the set of points in E at which f has at least two different finite derived numbers.

The sets X_1 and X_2 are easily shown to be denumerable and the set X_3 may be analyzed in exactly the same way as the set of points in the preceding theorem. We split X_3 into sequences of sets of the form X_{pqr} as in that theorem and obtain that $(S_0)\text{-}\mu_f$ vanishes on each such set. (We omit the details which merely repeat the construction used previously.) It follows now that the set X has $(S_0)\text{-}\mu_f$ zero as required.

The technical tool that provides the proof in each of these theorems is the agreement of the measures

$$(S_0)\text{-}\mu f \quad \text{and} \quad (S_\infty)\text{-}\mu f$$

on any set on which f is continuous and VBG∗∗. This same feature may be used in more general situations. In the next theorem one possible version is given.

(44.3) THEOREM. Let S be a local system that is filtering and let f be a real function. Suppose that f is S-variationally equivalent on a set E to a continuous function g that is VBG∗ on E, i.e.

$$V (\Delta f - \Delta g, S; E) = 0.$$

Then the (S)-derivative of f exists as a finite number at almost every point of E.

PROOF. Because of the variational equivalence and the fact that S is filtering we must have the agreement of the measures

$$(S)\text{-}\mu f = (S)\text{-}\mu g \quad \text{and} \quad (S^*)\text{-}\mu f = (S^*)\text{-}\mu g$$

on each subset of E. But the fact that g is continuous and has VBG∗ on the set E requires

$$(S_0)\text{-}\mu g = (S_\infty)\text{-}\mu g = (S)\text{-}\mu g = (S^*)\text{-}\mu g$$

on this set. Putting these together gives that
$$(S)\text{-}\mu_f = (S^*)\text{-}\mu_f$$
on E. This same agreement is available if f is replaced by a function of the form $f(x) + px$ for any real number p.

These observations allow us to prove the present theorem in precisely the same way that theorem (44.2) was proved.

§45. Absolute continuity.

Classically the notion of a function that is absolutely continuous arises in an ϵ, δ definition. The importance of this notion in integration theory was quickly recognized and nowadays it is the measure-theoretic version that many students will first see. We shall reproduce here the classical definition.

(45.1) DEFINITION. A function f is said to be <u>absolutely continuous</u> on an interval [a, b] if for every $\epsilon > 0$ there is a $\delta > 0$ so that for every sequence $\{I_1, I_2, \ldots, I_k\}$ of nonoverlapping subintervals of [a, b], if

$$\sum_{i=1}^{k} |I_i| < \delta \quad \text{then} \quad \sum_{i=1}^{k} |\Delta f(I_i)| < \epsilon.$$

This definition permits a restatement in terms of the variational measures and hence a generalization to more general settings.

(45.2) THEOREM. Let f be a continuous function on an interval [a, b]. Then f is absolutely continuous on [a, b] if and only if f has bounded variation on [a, b] and either of the three equivalent conditions holds:
(i) $(S_0)\text{-}\mu_f$ vanishes on each subset N of [a, b] of measure zero.
(ii) $|f[N]|$ vanishes on each subset N of [a, b] of measure zero.
(iii) $(S_0)\text{-}\mu_f (\{ x \in [a, b] : f'(x) = \pm\infty \}) = 0$.

PROOF. The first characterization (i) is classical since this measure is, for continuous functions f of bounded variation, the Lebesgue-Stieltjes measure generated by the total variation of f. The second (ii) then follows from (43.3) since the vanishing of $|f[N]|$ and $(S_0)\text{-}\mu_f (N)$ are equivalent.

For (iii) then partition the real line into the three sets X_1 where f has a finite derivative, X_2 where f has an infinite derivative, and X_3 where f has no derivative finite or infinite. By (44.2) we know that the measure $(S_0)\text{-}\mu_f$ vanishes on X_3. By (42.1) we know that $(S_0)\text{-}\mu_f$ will vanish on any measure zero subset of X_1. By (44.1) we know that X_2 has measure zero. From these facts we deduce that (i) and (iii) are equivalent.

A general version of this follows similarly. For the definition of ACG* see Saks [209].

(45.3) THEOREM. Let f be a continuous function on an interval $[a, b]$. Then f is ACG∗ on a closed set $E \subset [a, b]$ if and only if f is VBG∗ on E and either of the three equivalent conditions holds:
(i) (S_o)-μ_f vanishes on each subset N of E of measure zero.
(ii) $|f[N]|^e$ vanishes on each subset N of E of measure zero.
(iii) (S_o)-μ_f $(\{ x \in E : f'(x) = \pm \infty \}) = 0$.

PROOF. The characterization (ii) is classical; see the material in Chapter VII of Saks [209]. The remaining characterizations follow in exactly the same manner as in the proof of the preceding theorem.

§46. De la Vallée-Poussin decomposition theorem.

The relationship that exists between a function of bounded variation and its derivative was much explored in the early years of this century. Lebesgue showed that the total variation of a function f is related to the integral of the derivative by the inequality

$$\mathrm{Var}\,(\Delta f, [a, b]) \geq \int_a^b |f'(x)|\, dx.$$

(Here the function f' is considered to be defined everywhere by setting it to be zero at the points where the derivative does not exist.) This was sharpened by de la Vallée-Poussin to the precise identity

$$\mu_f(X) = \mu_f\left(\{ x \in X : f'(x) = \pm \infty \}\right) + \int_X |f'(x)|\, dx$$

holding for the total variation measure μ_f of a continuous function f of bounded variation and for any bounded Borel set X.

In this section we collect a number of results that are directly related to considerations of this kind.

The first result is an elementary computation showing that the measures (S)-μ_f have this same type of relationship for exact (S)-derivatives as appears in the above results.

(46.1) THEOREM. Let S be a filtering system and suppose that everywhere on a Borel set Y there is an exact S-derivative (S)-$D f(x) = g(x)$ where g is Lebesgue measurable on Y. Then

$$(S) - \mu_f(Y) = \int_Y |g(x)|\, dx.$$

PROOF. We will prove firstly that, under the hypotheses of the theorem, if the inequality
$$0 \leq a < |g(x)| < b$$
holds at every point of the set Y then
$$a\,|Y| \leq (S)\text{-}\mu_f(Y) \leq b\,|Y|.$$
To see this let C denote the collection of intervals
$$C = \{ I : a\,|I| < |\Delta f(I)| < b\,|I| \}$$

Since S is filtering, C is an S-cover of the set Y. Thus, if C_1 denotes an arbitrary S-cover of Y, we must have (using again the fact that S is filtering) the inequalities
$$a |Y| \leq a \, \text{Var}(\Delta x, C_1 \cap C) \leq \text{Var}(\Delta f, C_1 \cap C) \leq \text{Var}(\Delta f, C)$$
and
$$(S)\text{-}\mu_f(Y) \leq \text{Var}(\Delta f, C_1 \cap C) \leq b \, \text{Var}(\Delta x, C_1 \cap C) \leq b \, \text{Var}(\Delta x, C).$$
From these inequalities the result
$$a |Y| \leq (S)\text{-}\mu_f(Y) \leq b |Y|.$$
now follows directly.

Standard measure-theoretic arguments now are available to complete the proof.

If we apply these results to the ordinary derivatives (S_0-derivatives) and the measures $(S_0)\text{-}\mu_f$ we obtain the following versions of the results of Lebesgue and de la Vallée-Poussin.

(46.2) THEOREM. Let D_f denote the set of points where the function f has a finite derivative and let Y be an arbitrary Borel set. Then
$$(S_0)\text{-}\mu_f(Y) \geq \int_{Y \cap D_f} |f'(x)| \, dx.$$

PROOF. This follows directly from the preceding theorem since the set D_f is a Borel set (see e.g. Bruckner [33, p.228]).

(46.3) THEOREM. Let f be a continuous function, let D_f denote its set of points of finite differentiability, and let E denote a Borel set on which f is given to be VBG∗. Then the measure $(S_0)\text{-}\mu_f(E)$ may be given by the expression
$$(S_0)\text{-}\mu_f\Big(\{ x \in E : f'(x) = \pm \infty \}\Big) + \int_{E \cap D_f} |f'(x)| \, dx.$$

PROOF. Such a set E may be partitioned into three sets E_1, E_2, and E_3 where the function has a finite derivative, has an infinite derivative, and finally where the function has no derivative finite or infinite. By theorem (44.2) the last of these three sets has zero $(S_0)\text{-}\mu_f$ measure. What remains is precisely the decomposition presented in this theorem.

(46.4) Example. By way of an application of this theorem note that a function f that is continuous and given to be VBG∗ on a set E will be ACG∗ on E if and only if
$$(S_0)\text{-}\mu_f(\{ x \in E : f'(x) = \pm \infty \}) = 0.$$
In particular if f fails to be ACG∗ on E then the set of points in E at which $f'(x) = \pm \infty$ must be nondenumerable.

§47. Singular functions.

The classical notion of a singular function can be given a definition similar to the usual variational definitions for functions of bounded variation or absolutely continuous functions. We reproduce here that definition.

(47.1) DEFINITION. A function f defined on an interval $[a,b]$ is said to be **singular** if it has bounded variation there and if for every $\epsilon > 0$ there is a sequence $\{I_1, I_2, \ldots, I_k\}$ of nonoverlapping subintervals of $[a,b]$ for which

$$\sum_{i=1}^{k} |I_i| < \epsilon$$

and such that whenever $\{J_1, J_2, \ldots, J_m\}$ is a sequence of nonoverlapping subintervals of $[a,b]$ that do not overlap any interval in the sequence $\{I_i\}$ it is the case that

$$\sum_{i=1}^{m} |\Delta f(J_i)| < \epsilon .$$

This definition permits a restatement in terms of the variational measures and hence a generalization to more general settings. Let f be a function of bounded variation on an interval $[a,b]$. Then f is singular if and only if either of the two equivalent conditions holds:
(i) $f'(x) = 0$ a.e. in $[a,b]$.
(ii) (S_0)-μ_f is concentrated on a subset of $[a,b]$ of measure zero.

These characterizations suggest alternative ways of defining the concept, and indeed in general measure theory it is the characterization (ii) that is normally used. Yet a further way, which does not have the disadvantage of being restricted to functions of bounded variation, and which permits a variety of generalizations, arises directly from the variation itself. In this section we offer this definition and explore some of the implications.

(47.2) DEFINITION. A function f will be said to be **singular** on a set E provided the S_0-variation of the interval function $\sqrt{|\Delta f \Delta x|}$ vanishes on E, i.e. provided

$$V(\sqrt{|\Delta f \cdot \Delta x|}, S_0; E) = 0 .$$

Here $\sqrt{|\Delta f \cdot \Delta x|}$ is the interval function

$$I = [a,b] \to \sqrt{|\Delta f(I)| \cdot |I|} = \sqrt{|f(b)-f(a)|(b-a)} .$$

The reason this might be considered a plausable definition of the notion of singular function is that such a function f must have its growth somehow occuring on sets of small measure, in order for this variation to vanish.

Without proofs we sketch the main consequences of the definition. Each of these may be established by the methods studied in this chapter.

(47.3) For any function f and any set E one has
$$V(\sqrt{|\Delta f \cdot \Delta x|}, S_0; E) \leq \sqrt{V(\Delta f, S_0; E)} \sqrt{|E|}.$$

(47.4) Let f be a function that is everywhere differentiable on a Borel set E. Then
$$V(\sqrt{|\Delta f \cdot \Delta x|}, S_0; E) = \int_E \sqrt{|f'(x)|}\, dx.$$

(47.5) A function f is singular on a set E if any of the following hold:
(i) f has zero variation on E.
(ii) $f'(x) = 0$ everywhere on E.
(iii) E has measure zero and f has σ-finite variation on E.
(iv) $f'(x) = 0$ a.e. on E and f has σ-finite variation on E.

(47.6) Let f be a function of bounded variation on an interval $[a, b]$ and extended to all of $\mathbb{R}$ by writing $f(x) = f(a)$ if $x < a$ and $f(x) = f(b)$ if $x > b$. Then f is singular in the sense of definition (47.1) if and only if f is singular on $[a, b]$ (in the sense of definition (47.2)).

We prove that singular functions must have almost everywhere vanishing derivatives.

(47.7) THEOREM. Let the function f be singular on a set E. Then $f'(x) = 0$ a.e. on E.

PROOF. For any integer n let E_n denote the set of points in E at which the function f has a finite or infinite derived number $d(f, x)$ with $|d(f, x)| > 1/n$. Then the collection of intervals
$$C = \{I : |\Delta f(I)| > |I|/n\}$$
is a S_∞-cover of E_n. This gives the inequality
$$\text{Var}(\Delta x, C) \leq \sqrt{n}\, \text{Var}(\sqrt{|\Delta f \Delta x|}, C).$$
Thus if C_1 is any S_0-cover of E we have, using the properties of such covers,
$$|E_n| \leq \text{Var}(\Delta x, C \cap C_1) \leq$$
$$\sqrt{n}\, \text{Var}(\sqrt{|\Delta f \Delta x|}, C \cap C_1) \leq \sqrt{n}\, \text{Var}(\sqrt{|\Delta f \Delta x|}, C_1).$$
Since the extreme right member of this inequality may be made as small as we please by a suitable choice of C_1 it follows that $|E_n| = 0$.

From this we may conclude that at almost every point $x \in E$ the function f has no derived number, finite or infinite, other than zero. Consequently $f'(x)$ exists and vanishes a.e. in E.

Somewhat more generally we may define a notion of mutual singularity for any pair of functions.

(47.8) DEFINITION. Two functions f and g defined on an interval $[a,b]$ are said to be <u>mutually singular</u> on a set E provided
$$V(\sqrt{|\Delta f \Delta g|}, S_0; E) = 0.$$

Note that should f and g be mutually singular on each member of a sequence of sets $\{E_i\}$ then f and g are mutually singular on any subset of the union of the sequence. Thus, for any fixed pair of functions f and g, the collection of sets on which they are mutually singular is a σ-ideal. If the functions are continuous, or at least bounded and not discontinuous from the same side at any point, then this σ-ideal includes all denumerable sets. We begin with a simple statement that gives a particular instance of mutual singularity. For the remainder of this section we use μ_f to denote the S_0-variational measure (Stieltjes measure) for the function f and the variational statements refer to this type of variation..

(47.9) LEMMA. If the function f has zero variation on a set E and g has σ-finite variation on that set then f and g are mutually singular on E.

PROOF. We may suppose that $\mu_g(E) < +\infty$. Let $\varepsilon > 0$ and choose full covers C_1 and C_2 of E so that
$$\text{Var}(\Delta g, C_1) < \mu_g(E) + 1 \text{ and } \text{Var}(\Delta f, C_2) < \varepsilon.$$
Then, using the Cauchy-Schwartz inequality in an obvious manner, we obtain
$$V(\sqrt{|\Delta f \Delta g|}, S_0; E) \leq \text{Var}(\sqrt{|\Delta f \Delta g|}, C_1 \cap C_2) \leq \sqrt{\varepsilon[\mu_g(E) + 1]}.$$
From this, since ε is an arbitrary positive number, we see that f and g are mutually singular on E.

Our next result gives a characterization of the notion of mutually singular functions in the case that the functions have finite variation.

(47.10) LEMMA. Let the functions f and g have finite variation on a set E. Then the following assertions are equivalent:
(a) f and g are mutually singular on E.
(b) for every $\varepsilon > 0$ there is a S_0-cover C of E such that any partition π contained in C may be decomposed so that
$$\pi = \pi_1 \cup \pi_2, \pi_1 \cap \pi_2 = \emptyset,$$
$$\sum_{I \in \pi_1} |\Delta f(I)| < \varepsilon, \text{ and } \sum_{I \in \pi_2} |g(I)| < \varepsilon.$$

PROOF. Suppose that f and g are mutually singular on E and let $\varepsilon > 0$. Choose a

S_0-cover C of E so that
$$\text{Var}(\sqrt{|\Delta f \Delta g|}, C) < \varepsilon.$$
For any partition π contained in C we shall write
$$\pi_1 = \{I \in \pi : |\Delta f(I)| < |\Delta g(I)|\}$$
and
$$\pi_2 = \{I \in \pi : |\Delta f(I)| \geq |\Delta g(I)|\}.$$
We observe that
$$\sum_{I \in \pi_1} |\Delta f(I)| \leq \sum_{I \in \pi_1} \sqrt{|\Delta f(I) \Delta g(I)|} < \varepsilon$$
and
$$\sum_{I \in \pi_2} |\Delta g(I)| \leq \sum_{I \in \pi_2} \sqrt{|\Delta f(I) \Delta g(I)|} < \varepsilon.$$
This gives the implication (a) → (b).

In the other direction let $\varepsilon > 0$ and let us choose S_0-covers C_1, C_2, and C_3 of E so that
$$\text{Var}(\Delta f, C_1) \leq V(\Delta f, S_0; E) + 1$$
$$\text{Var}(\Delta g, C_2) \leq V(\Delta g, S_0; E) + 1$$
and so that C_3 satisfies the hypotheses of assertion (b). Then for any partition π contained in $C_4 = C_1 \cap C_2 \cap C_3$ we may form a decomposition $\pi = \pi_1 \cup \pi_2$ so that
$$\sum_{I \in \pi_1} |\Delta f(I)| < \varepsilon \quad \text{and} \quad \sum_{I \in \pi_2} |\Delta g(I)| < \varepsilon.$$
Then an elementary application of the Cauchy-Schwartz inequality will give
$$\sum_{I \in \pi} \sqrt{|\Delta f(I) \Delta g(I)|} \leq \sqrt{\varepsilon[V(\Delta f, S_0; E) + 1]} + \sqrt{\varepsilon[V(\Delta g, S_0; E) + 1]}.$$
Since this applies to any partition π from C_4 and since ε is an arbitrary positive number the variation
$$V(\sqrt{|\Delta f \Delta g|}, S_0; E)$$
must vanish as required to establish that f and g are mutually singular on E.

For functions of bounded variation on the entire interval $[a, b]$ this is related to a similar result of Garg [97, Theorem 2, p.55]. We obtain this as a corollary by using the fact that every S_0-cover of an interval $[a, b]$ contains a partition of the interval. Note that the condition here is closely related to the pattern in the original defintion (47.1) for singular functions.

(47.11) COROLLARY. Let the functions f and g have bounded variation on the interval $[a, b]$. Then the following assertions are equivalent:
(a) f and g are mutually singular on $[a, b]$.

(b) for every $\varepsilon > 0$ there is a partition π of $[a, b]$ that may be decomposed so that

$$\pi = \pi_1 \cup \pi_2, \quad \pi_1 \cap \pi_2 = \emptyset,$$

$$\sum_{I \in \pi_1} |f(I)| > \mu_f([a, b]) - \varepsilon,$$

and

$$\sum_{I \in \pi_2} |g(I)| > \mu_g([a, b]) - \varepsilon.$$

PROOF. The condition (b) of the lemma may be seen to be equivalent, under these additional hypotheses, to the present condition (b). Thus if C is a S_0-cover of $[a, b]$ with the properties expressed in (b) of the lemma then there exists a partition π of $[a, b]$ contained in C such that

$$\sum_{I \in \pi} |\Delta f(I)| > \mu_f([a, b]) - \varepsilon.$$

By passing to a finer partition we may also assume that

$$\sum_{I \in \pi} |\Delta g(I)| > \mu_g([a, b]) - \varepsilon.$$

Now by the lemma there is a decomposition π_1, π_2 of π so that

$$\sum_{I \in \pi_2} |\Delta f(I)| < \varepsilon \quad \text{and} \quad \sum_{I \in \pi_1} |g(I)| < \varepsilon.$$

These inequalities then give

$$\sum_{I \in \pi_1} |\Delta f(I)| > \mu_f([a, b]) - 2\varepsilon,$$

and

$$\sum_{I \in \pi_2} |\Delta g(I)| > \mu_f([a, b]) - 2\varepsilon,$$

which supplies the assertion (b) of the corollary.

In the opposite direction if the condition (b) of the corollary holds for some ε, π, π_1 and π_2 then we define C_1 to be the collection of all intervals that are subintervals of some member of the parition π. Let C_2 and C_3 be chosen as S_0-covers of $[a, b]$ in such a way that

$$\text{Var}(\Delta f, C_1) < \mu_f([a, b]) + \varepsilon \quad \text{and} \quad \text{Var}(\Delta g, C_1) < \mu_g([a, b]) + \varepsilon.$$

Now let $C = C_1 \cap C_2 \cap C_3$. This is a S_0-cover of $[a, b]$ and any partition π_3 contained in C may be decomposed according to the decomposition of the original parition π: thus

$$\pi_4 = \{ I \in \pi_3 : I \text{ a subinterval of a } J \in \pi_1 \}$$

and

$$\pi_5 = \{ I \in \pi_3 : I \text{ a subinterval of a } J \in \pi_2 \}.$$

Certainly $\pi_4 \cap \pi_5 = \emptyset$, $\pi_3 = \pi_4 \cap \pi_5$, and the inequalitiies

$$\sum_{I \in \pi_3} |f(I)| < 2\epsilon, \text{ and } \sum_{I \in \pi_4} |g(I)| < 2\epsilon$$

are readily established. This shows that the condition (b) of the lemma may be obtained directly from the condition (b) of the corollary and so the proof is complete.

We may now characterize in another way the mutual singularity of functions in the case that the functions have finite variation. This extends easily to sets on which the variation is σ-finite and so, in particular, may be applied to functions that are VBG$_*$.

(47.12) THEOREM. Let the functions f and g have finite variation on a set E. Then in order that f and g be mutually singular on the set E the following three conditions are necessary and sufficient:
(i) at each point $x \in E$, $\lim_{h \to 0} [f(x+h) - f(x)][g(x+h) - g(x)] = 0$.
(ii) $f'(x)g'(x) = 0$ a.e. in E.
(iii) the set of points in E at which f and g are continuous with $f'(x) = \pm\infty$ and $g'(x) = \pm\infty$ is the union of a set of μ_f-measure zero and a set of μ_g-measure zero.

PROOF. To begin let us suppose that the functions f and g satisfy the conditions asserted. Then we observe that, because μ_f and μ_g are finite on E, there are at most denumerably many points of discontinuity of either function in E. If E_0 denotes this set of points then, by (i), we check easily that $V(\sqrt{|\Delta f \Delta g|}, S_0; E_0) = 0$.

Let E_1 and E_2 denote respectively the sets of points in x at which $f'(x) = 0$ and $g'(x) = 0$. By lemma (47.9) we must have

$$V(\sqrt{|\Delta f \Delta g|}, S_0; E_1) = V(\sqrt{|\Delta f \Delta g|}, S_0; E_2) = 0.$$

Now let E_3 and E_4 denote respectively the sets of points x in E at which f is continuous but $f'(x)$ fails to exist (finitely or infinitely) and at which g is continuous and $g'(x)$ fails to exist. By (44.2) we must have $\mu_f(E_3) = \mu_g(E_4) = 0$ and thus, again by (47.9),

$$V(\sqrt{|\Delta f \Delta g|}, S_0; E_3) = V(\sqrt{|\Delta f \Delta g|}, S_0; E_4) = 0.$$

Finally if E_5 denotes the set of points at which f and g are continuous with $f'(x) = \pm\infty$ and $g'(x) = \pm\infty$ then the condition (iii) of the theorem together with yet a further application of (47.9) supplies immediately

$$V(\sqrt{|\Delta f \Delta g|}, S_0; E_5) = 0.$$

This expresses E as a union of sets on each of which f and g are mutually singular and thus the first part of the theorem is proved.

In the opposite direction let us suppose that

$$V(\sqrt{|\Delta f \Delta g|}, S_0; E) = 0,$$

and let us verify the three conditions (i), (ii), and (iii). The first condition is obvious since for each $x \in E$,

$$V(\sqrt{|\Delta f \Delta g|}, S_0; \{x\}) = 0.$$

By (44.2) we know that $f'(x)$ and $g'(x)$ exist finitely a.e.. The set of points A at which $f'(x)g'(x)$ exists but differs from zero may be written as a union of sets $\{A_n\}$ where

$$A_n = \left\{ x \in E : f'(x), g'(x) \text{ exist}, |f'(x)g'(x)| > \frac{1}{n} \right\}.$$

The collection of intervals

$$C_1 = \left\{ I : |\Delta f(I) \Delta g(I)| > \frac{1}{n} |I|^2 \right\}$$

is a S_0-cover of A_n and so for any S_0-cover C_2 of E we have that $C = C_1 \cap C_2$ is a S_0-cover of A_n and hence

$$|A_n| \leq \text{Var}(\Delta x, C) \leq n \text{Var}(\sqrt{|\Delta f \Delta g|}, C) \leq n \text{Var}(\sqrt{|\Delta f \Delta g|}, C_2).$$

Since this may be chosen arbitrarily small the measure of each A_n vanishes and thus A is a set of measure zero as required to prove assertion (ii).

Finally it remains to prove (iii). Let B denote the set of points x in E at which both f and g are continuous with $|f'(x)| = |g'(x)| = \pm \infty$. We let C be the set of points

$$\left\{ x \in B : \limsup_{h \to 0} \left| \frac{f(x+h) - f(x)}{g(x+h) - g(x)} \right| < +\infty \right\}.$$

and write

$$C_n = \left\{ x \in C : \limsup_{h \to 0} \left| \frac{f(x+h) - f(x)}{g(x+h) - g(x)} \right| < n \right\}.$$

The collection C_1 of intervals I for which $|\Delta f(I)| < n |\Delta g(I)|$ is clearly a S_0-cover of the set C_n. Setting again $C = C_1 \cap C_2$ this gives the estimate

$$\text{Var}(\Delta f, C) \leq \sqrt{n} \, \text{Var}(\sqrt{|\Delta f \Delta g|}, C_2)$$

from which we may derive, repeating the same arguments as used above, that μ_f vanishes on each C_n. Thus $\mu_f(C) = 0$.

We now show that $\mu_g(B \setminus C) = 0$. For any $\varepsilon > 0$ the collection

$$C_1 = \{ I : |\Delta g(I)| < \varepsilon |\Delta f(I)| \}$$

is a S_∞-cover of $B \setminus C$, for if not then some point x in $B \setminus C$ would belong to a set C_n with $n > 1/\varepsilon$. Thus we must have, with the same notation,

$$V(\Delta g, S_\infty; B \setminus C) \leq \text{Var}(\Delta g, C) \leq \sqrt{\varepsilon} \, \text{Var}(\sqrt{|\Delta f \Delta g|}, C_2).$$

By (41.6) this requires that $\mu_g(B \setminus C) = 0$ as required and the theorem is proved.

CHAPTER FIVE
MONOTONICITY

§48. Introduction.

There are in the literature many theorems whose conclusion is that a function is monotonic on some set. Of course results of this type have important applications, and consequently the basic monotonicity theorems have been subjected to intensive study and extensive generalization. Our purpose in this chapter is to present a somewhat systematic presentation of several classes of monotonicity theorems. These are of interest in their own right and are also needed in later chapters. In particular many theorems in Chapter six which assert relations that hold among certain generalized derivations will be seen to arise directly from an appropriate monotonicity theorem.

There are three different types of monotonicity result that we wish to present and that we shall discuss under the headings, <u>local monotonicity</u>, <u>relative monotonicity</u>, and <u>global monotonicity</u>.

A function is locally increasing at a point x if there is some neighbourhood $(x - \delta, x + \delta)$ so that
$$f(y) < f(x) < f(y') \quad \text{if} \quad x - \delta < y < x < y' < x + \delta.$$
Various questions arise from this concept and numerous generalizations are possible. These we consider as local monotonicity questions and will discuss in sections 49 and 50.

A function is increasing (globally) if whenever $x < y$ it is the case that $f(x) < f(y)$. The first monotonicity theorem that we all learn is that from the elementary calculus stating that an everywhere differentiable function whose derivative is positive is increasing. Such a result, whose conclusion is that a function is monotonic on the entire real line, we call a global monotonicity theorem; a number of global monotonicity theorems are presented in sections 54, 55, and 56.

In many instances in which the hypotheses are much weaker one can conclude only that a given function is monotonic on certain sets, but not globally. Such a theorem we call a relative monotonicity theorem. For example it is easy to show that a function f, that has everywhere a positive Dini derivative $D^+ f(x) > 0$, need not be monotonic, but is monotonic on each of a sequence of sets $\{E_n\}$ whose union is the entire real line. Also related to this are classical notions such as VBG∗, ACG∗ etc. wherein the growth of a function relative to a sequence of sets is specified. In sections 51, 52, and 53 some theorems of this type are given.

§49. Local monotonicity.

In this section we consider the notion of increase or decrease of a function at a point. The simplest such concept may be formulated as follows: we say that f is <u>bilaterally increasing</u> (i.e. strictly) at a point x provided there is a neighbourhood $(x - \delta, x + \delta)$ so that
$$f(y') < f(x) < f(y'') \quad \text{if} \quad x - \delta < y' < x < y'' < x + \delta.$$

Certainly if $\underline{D}f(x) > 0$ then f is bilaterally increasing at x, but the converse is not true.

This concept and its generalizations suggest several problems. The first of these is a simple global monotonicity theorem given by Heindl and Köhler [111].

(49.1) THEOREM. If a function f is bilaterally increasing at each point then it must be in fact increasing.

PROOF. The proof is an easy application of (16.6) or may be obtained by an obvious compactness argument.

One might ask too for the nature of the set of points x at which an arbitrary function f is bilaterally increasing (see Real Analysis Exchange 2 (1976), p.77). Note that the set of points of one-sided increase may be entirely arbitrary: let E be any set and write $f(x) = \pi/2$ $(x \notin E)$ and $f(x) = \arctan x$ $(x \in E)$. Thus f is increasing on the right, precisely at the points of the set E.

(49.2) THEOREM. The set of points x at which an arbitrary function f is bilaterally increasing is a $G_{\delta\sigma}$.

PROOF. If E is the set of points of bilateral increase then, for each $x \in E$, the set
$$S_x = \{ y : \frac{f(y) - f(x)}{y - x} > 0 \}$$
is a neighbourhood of x. Thus there is a guage δ on E so that
$$\delta(x) = \sup \{ t : (x - t, x + t) \subset S_x \}.$$
Define the sets
$$E_{nm} = \{ x \in E : \delta(x) > \frac{1}{n} \} \cap [\frac{m}{n}, \frac{m+1}{n}) \}$$
for $n = 1, 2, \ldots, m = 0, \pm 1, \pm 2, \ldots$, whose union is evidently E. Each set E_{nm} may be seen to be a G_δ from the following observation. If x is a point of bilateral accumulation of E_{nm} then x must in fact be in E_{nm}. For if $y_k \to x$ $(y_k < x)$ and $z_k \to x$ $(z_k > x)$ are sequences in E_{nm} then for every pair of numbers y' and y'' with
$$x - \frac{1}{n} < y' < x < y'' < x + \frac{1}{n}$$
one has $f(y') < f(x) < f(y'')$, since for some index k,
$$y_k - \frac{1}{n} < y' < y_k < x < y_k + \frac{1}{n}$$
and
$$z_k + \frac{1}{n} > y'' > z_k > x > z_k - \frac{1}{n}.$$
This means that each set E_{nm} is a closed set less the denumerable set of points that are isolated on one side, and hence must be a G_δ. It then follows that E is a $G_{\delta\sigma}$ as required.

§50. Generalized local monotonicity.

This notion of local monotonicity extends in a natural way to the more general setting of a local system. We give the definition and a few selected results.

(50.1) DEFINITION. Let S be a local system. A function f is said to be S-<u>increasing</u> at a point x provided
$$\{y: y = x \text{ or } \frac{f(y) - f(x)}{y - x} > 0\} \in S(x).$$
In a similar way one would define S-decreasing, S-nonincreasing, and S-nondecreasing.

The above monotonicity theorem (49.1) extends immediately to this more general notion of local increase.

(50.2) THEOREM. Let S be a local system that is at each point bilateral and which satisfies an intersection condition of the form
$$S_x \cap S_y \cap [x, y] \neq \emptyset.$$
Then, if a function f is S-increasing at each point, it must in fact be increasing.

PROOF. This is a direct application of the covering lemma (16.6).

We turn now to some considerations of this notion that arise from density ideas. For the moment let us say that a function f is <u>approximately increasing</u> at a point x on the right (left) if the set
$$\{y: \frac{f(y) - f(x)}{y - x} > 0\}$$
has exterior density 1 on the right (left) at x, and that f is <u>weakly approximately increasing</u> there if this set has exterior <u>upper</u> density 1 in the same sense. Similarly one has approximate notions for decreasing, nonincreasing, and nondecreasing. Note that these notions arise also from definition (50.1) above using an appropriate density system. We have the following result.

(50.3) LEMMA. Let f be an arbitrary function. Then, at almost every point x, if f fails to be weakly approximately increasing (nondecreasing) on the right at x then f must be approximately nonincreasing (decreasing) on the left at x.

PROOF. Let E be a set of points x at which f fails to be weakly approximately increasing on the right at x. Then for each $x \in E$ the set
$$\{y: f(y) > f(x)\}$$
fails to have right exterior upper density 1 at x, which means that
$$S_x = \{y: f(x) \leq f(y)\}$$
has right interior lower density positive at x. Let E_p $(p > 0)$ denote the set of points x

at which S_x has this density exceeding p. We define a guage δ on E_p so that
$$|S_x \cap (x, x+t)|^i > pt \quad \text{for} \quad 0 < t < \delta(x),$$
and let $\{E_{pn}\}$ denote the partition of E_p induced by the guage δ.

We claim that at almost every point of a set E_{pn} the function f must be approximately nonincreasing on the left. Indeed almost every point x of E_{pn} is a point of exterior density 1 of that set as well as a point of weak approximate continuity of f (see (24.3)). Let x_0 be such a point: then E_{pn} has exterior density 1 at x_0 and there is a set A also with exterior density 1 at x_0, relative to which f is continuous at x_0. Let us show that for every $z \in E_{pn}$ ($z < x_0$) sufficiently close to x_0, $f(z) \geq f(x_0)$; this will show that f is approximately nonincreasing on the left at x_0 as required.

Choose δ_0 so that
$$|E_{pn} \cap (x_0 - t, x_0)|^e > (1 - \tfrac{p}{2})t \quad \text{if} \quad 0 < t < \delta_0,$$
and
$$|A \cap (x_0 - t, x_0)|^e > (1 - \tfrac{p}{2})t \quad \text{if} \quad 0 < t < \delta_0,$$
and let z be any point in E_{pn} with $x_0 - \delta_0 < z < x_0$. Now we produce sequences of points $\{z_n\}$ and $\{w_n\}$ so that $z_0 = z$, $z < z_n < x_0$, $z_n \to x_0$, $z_n \leq w_n \leq x_0$,
$$z_n \in E_{pn} \cap S_{z_{n-1}} \quad \text{and} \quad w_n \in S_{z_n} \cap A.$$
These sequences must exist because of the density requirements on the sets A, E_{pn}, and S_{z_n}.

This gives
$$f(w_n) \leq f(z_n) \leq f(z_{n-1}) \leq f(z)$$
and since $f(w_n) \to f(x_0)$ we obtain $f(x_0) \leq f(z)$ as asserted above. This completes the proof.

From this lemma we may deduce the following result of Khintchine [136] from 1924 (cf. Davies [57]). For convenience let us express it in this language. A function f is <u>approximately constant</u> at a point x if the level $f^{-1}(\{f(x)\})$ has (exterior) density 1 at the point x. A function is <u>approximately oscillatory</u> at x if it is both weak approximately increasing and weak approximately decreasing (bilaterally) at that point. It can be shown that at points of type (i), (ii), or (iii) in the statement of the theorem the function f is a.e. approximately differentiable; thus any measurable function f is, at almost every point, either approximately differentiable or else approximately oscillatory. From this we can also conclude that such a function must have at almost every point an approximate derivative or else $\pm\infty$ are essential derived numbers.

(50.4) THEOREM. Let f be a measurable function. Then almost every point x falls into one of the following classes:
(i) f is approximately constant at x,
(ii) f is approximately increasing at x,
(iii) f is approximately decreasing at x, or
(iv) f is approximately oscillatory at x.

PROOF. To begin, let A denote the set of points x at which the level
$$L_c = \{ y : f(y) = c \}$$
has positive measure. There are only denumerably many such values of c and at the density points of each level f is approximately constant. Thus a.e. point of A may be placed in the class (i), leaving us to consider only the points x at which the set $\{ y : f(y) = f(x) \}$ has measure zero.

At these points now we apply lemma (50.3), remembering that f is measurable so that we need not distinguish between interior and exterior density. For almost every point x not in A if the set
$$S_x = \{ y : f(y) > f(x) \}$$
has positive lower density on the right at x, then the set
$$T_x = \{ y : f(y) < f(x) \}$$
has density 1 on the left at x. This fact, together with its variations replacing left by right and changing the inequalities, evidently supplies the proof.

(50.5) Example. A result that is closely related to the above considerations appears in the article of Khintchine [136]. This article suggests that for continuous functions a porosity condition may replace the density conditions. If fact the article contains an error and porosity conditions do not give any information about the approximate differentiability of a function. One can give an example to illustrate this.

There exists a continuous function f defined on the interval [0,1], and a set $P \subset [0,1]$ of positive measure such that f does not have an approximate derivative at any point of P and, for each $x \in P$, both of the sets
$$S_x = \{ y : y = x \text{ or } f(y) > f(x) \}$$
and
$$T_x = \{ y : y = x \text{ or } f(y) < f(x) \}$$
are nonporous at x.

The construction is given in Bruckner, Laczkovich, Petruska and Thomson [41]. From this same construction we may conclude that the existence of a generalized derivative, even for a nonporous system, need not imply the existence of the approximate derivative. This shows that porosity conditions may not substitute for intersection conditions, or for density conditions in theorems of this type: there exists a continuous function f defined on the interval [0,1], a set $P \subset [0,1]$ of positive measure, and a local system **S** such that for each $x \in P$ and each $S \in S(x)$, S is nonporous at x, the derivative $(S)\text{-}Df(x)$ exists and vanishes at every point $x \in P$, and yet f does not have an approximate derivative at any point of P.

§51. Relative monotonicity.

We turn our attention now to some results asserting the relative monotonicity behaviour of functions. Each of these results gives that, under appropriate assumptions, there is a denumerable decomposition of some set X into a sequence of sets $\{X_n\}$ in such a way that a given function is monotonic on each member of the sequence.

Our basic result in this section asserts that, for a derivation taken with respect to a system that has an appropriate intersection property, a useful relative monotonicity theorem is available. For the special systems that express ordinary bilateral and unilateral derivation somewhat sharper intersection conditions are available and this gives rise to more detailed relative monotonicity results; see (51.4) and the subsequent comments. More subtle versions are given in a later section.

(51.1) LEMMA. Let f be an abitrary function and let S be a local system that satisfies an intersection condition of the form
$$S_x \cap S_y \cap [x, y] \neq \emptyset.$$
If everywhere on a set X the inequality $(S)\text{-}\underline{D} f(x) > r$ holds, then there is a denumerable partition $\{X_n\}$ of the set X so that the function $f_r(x) = f(x) - rx$ is strictly increasing on each set X_n.

PROOF. Define for each point $x \in X$ the sets
$$S_x = \{ y : y = x \text{ or } \frac{f(y) - f(x)}{y - x} > r \}$$
and observe that, because of the hypotheses, each $S_x \in S(x)$. As we are assuming the intersection condition stated, this choice from S induces a denumerable partition $\{X_n\}$ of the set X in such a way that whenever x, y ($x < y$) belong to the same member of the partition the intersection
$$S_x \cap S_y \cap [x, y]$$
contains some point z. One considers separately the situations $z = x$, $z = y$, and $x < z < y$. In each case it is easy to verify that this requires $f_r(x) < f_r(y)$ which is the desired conclusion of the lemma.

Note that one can state somewhat more. The partition also has the property that for points x_1 and y_1 in the closure of X_n relative to X the same nonempty intersection argument is available by taking points x and y in X_n sufficiently close to x_1 and y_1 respectively. Thus one can prove that if x_1 and y_1 ($x_1 < y_1$) are in $X \cap \overline{X}_n$ with x_1 not isolated on the right there and y_1 not isolated on the left then one has
$$\frac{f(y_1) - f(x_1)}{y_1 - x_1} > r.$$
This is also of some use.

(51.2) COROLLARY. Let f be an arbitrary function and let S be a local system that is filtering and satisfies an intersection condition of the form

$$S_x \cap S_y \cap [x,y] \neq \emptyset.$$
If everywhere on a set X the inequalities
$$-\infty < (S)\text{-}\underline{D}\,f(x) \leq (S)\text{-}\overline{D}\,f(x) < +\infty,$$
hold, then there is a denumerable partition $\{X_n\}$ of the set X so that f satisfies a Lipschitz condition
$$|f(x) - f(y)| \leq M_n |x - y|$$
on each set $X \cap \overline{X}_n$.

PROOF. (Note that a continuous function f that satisfies such a Lipschitz condition on each of the sets X_n will be ACG on X.) For each integer m define the set,
$$Y_m = \{x \in X : -m < (S)\text{-}\underline{D}\,f(x) \leq (S)\text{-}\overline{D}\,f(x) < m\}$$
and argue as in the preceding lemma to produce a sequence of sets, $\{Y_{nm}\}$ so that the inequality
$$\left|\frac{f(y) - f(x)}{y - x}\right| \leq m$$
holds for x, y in the same set Y_{nm}. This extends to the closures taken in X, again by the same arguments as indicated above. To conclude, one merely relabels the sequence Y_{nm} as a singly indexed sequence $\{X_n\}$.

(51.3) COROLLARY. Let f be a continuous function and let S be a local system that satisfies an intersection condition of the form
$$S_x \cap S_y \cap [x,y] \neq \emptyset.$$
If everywhere on a set X of the second category the inequality
$$(S)\text{-}\underline{D}\,f(x) > r$$
holds, then there is an interval (a,b) on which the function $f_r(x) = f(x) - rx$ is increasing.

PROOF. We apply the lemma to produce a partition $\{X_n\}$ of the set X so that on each set X_n the function f_r is increasing. If, contrary to the assertion of the corollary, there is no such interval (a,b) on which f_r is increasing then each of these sets X_n is nowhere dense. But this contradicts the fact that X itself is second category, and this contradiction proves the result.

The above results use the basic intersection condition,
$$S_x \cap S_y \cap [x,y] \neq \emptyset.$$
For specific systems usually a more detailed intersection condition may be given, and this may be used in an identical fashion to produce somewhat better results. By way of illustration we take the ordinary bilateral and unilateral derivations, which satisfy the above basic intersection condition, and show how, by identical arguments, we may improve upon the statements in the lemma and its corollaries.

The system S_0 that expresses ordinary limits and hence ordinary derivates has in fact the intersection condition
$$S_x \cap S_y \supset [x, y]. \qquad (*)$$
If we use this intersection condition we may obtain in an obvious manner the following results. Let us say, for the purposes only of this lemma, that a function f is increasing around a set E provided $x, y \in E$, and $x < z < y$ implies that $f(x) < f(z) < f(y)$. Then $(*)$ supplies this improvement of (51.1).

(51.4) LEMMA. Let f be a function and suppose that everywhere on a set X the inequality
$$\underline{D} f(x) > r$$
holds. Then there is a denumerable partition $\{X_n\}$ of the set X so that the function $f_r(x) = f(x) - rx$ is strictly increasing around each set X_n.

As a corollary we may obtain that a function f that has for every x in a set X either $\underline{D} f(x) > -\infty$ or $\overline{D} f(x) < +\infty$, must be VBG$*$ on the set X (cf. Saks [209, Theorem 10.1, p.234]). We obtain also that a function f that has both derivates finite,
$$-\infty < \underline{D} f(x) \leq \overline{D} f(x) < +\infty$$
everywhere on a set X permits a denumerable partition $\{X_n\}$ of the set X in such a way that f satisfies a strong Lipschitz condition,
$$\sup\nolimits_{z_1, z_2 \in [x, y]} |f(z_1) - f(z_2)| \leq M_n |x - y|$$
for $x, y \in \overline{X}_n \cap X$. This is just a sharper version of (51.2) above proved using the intersection condition $(*)$. Note that in particular if f is continuous this will require f to be ACG$*$ on X (cf. Saks [209, Theorem 10.5, p.235]).

For the Dini derivatives the basic intersection condition may be slightly improved to the assertion that
$$y \in S_x \cap S_y \quad \text{and} \quad S_x \supset [x, y]. \qquad (**)$$
This does not help in (51.1), so that we can conclude only for a function f, with $\underline{D}^+ f(x) > r$ everywhere on a set X, that f_r is increasing on each member X_n of a denumerable partition of X. This gives that a function f that has either $\underline{D}^+ f(x) > -\infty$ or $\overline{D}^+ f(x) < +\infty$ everywhere on a set X must be VBG on X (cf. Saks [209, Theorem 10.8, p.237]). We may however use the intersection condition $(**)$ to obtain that a function f that has both Dini derivatives on one side finite,
$$-\infty < \underline{D}^+ f(x) \leq \overline{D}^+ f(x) < +\infty$$
everywhere on a set X permits a denumerable partition $\{X_n\}$ of the set X in such a way that f satisfies a strong Lipschitz condition,
$$\sup\nolimits_{z_1, z_2 \in [x, y]} |f(z_1) - f(z_2)| \leq M_n |x - y|$$
for $x, y \in \overline{X}_n \cap X$. Note that, again, if f is continuous this will require f to be ACG$*$ on X (cf. Saks [209, Theorem 10.5, p.235]).

(51.5) Example. The basic results (51.1), (51.2), and (51.3) apply to the bilateral

approximate derivates. In particular, if a function f has
$$\underline{D}_{ap} f(x) > -\infty \quad \text{or} \quad \overline{D}_{ap} f(x) < +\infty$$
everywhere on a set X, then f must be VBG on X (cf. Saks [209, Theorem 10.8, p. 237]). For one-sided approximate derivates this intersection condition is not available and this result fails. However more delicate considerations may be applied; these appear in the next section.

§52. Further relative monotonicity theorems.

The basic monotonicity theorem of the preceding section and its variants are quite elementary. In this section we collect a few more specialized versions of this kind of theorem. The first of these asserts that for a continuous function a very weak kind of positive derivative assures some relative monotonicity results. Note that this lemma arises from a porosity assumption and not from an intersection condition.

(52.1) LEMMA. Let f be a continuous function, and let S be a local system so that, at each point x, if $S \in S(x)$ then S has right hand porosity less than 1 at x. Then, if $(S)\text{-}\underline{D} f(x) > r$ at every point x of a set X, there is a denumerable partition
$$X = N \cup X_1 \cup X_2 \cup X_3 \cup \ldots$$
of the set X such that N is first category and the function $f_r(x) = f(x) - rx$ is increasing on each set X_n.

PROOF. We may suppose that at every point $x \in X$ the inequality
$$(S) - \underline{D} f(x) > s > r$$
holds for some $s > r$ since, if not, the set X may be partitioned into a denumerable family within each member of which such an inequality is valid. Thus at each point $x \in X$ we define the set
$$S_x = \{ y : y = x \text{ or } \frac{f(y) - f(x)}{y - x} > s > r \}.$$
As each such set is in $S(x)$, S_x has right porosity at x less than 1. By forming a further partition we may suppose that this porosity is smaller than some $p < 1$. We define a guage δ on X in such a way that
$$\lambda (S_x, x, x + t) < pt \quad \text{for } 0 < t < \delta(x)$$
and use δ to induce a denumerable partition of the set X in the standard manner. Let us denote this partition as $\{Y_n\}$.

We further partition each set Y_n so that
$$Y_n = N_n \cup Y_{n1} \cup Y_{n2} \cup Y_{n3} \cup \ldots \qquad (*)$$
where N_n is nowhere dense and the set Y_n is dense in each of the intervals (c_{nk}, d_{nk}), where $c_{nk} = \inf Y_{nk}$, and $d_{nk} = \sup Y_{nk}$. This is done by the following construction: we let Z_n denote the set of points x in Y_n for which there is a positive number $\delta_1(x)$ such that Y_n is dense in the interval $(x-\delta_1(x), x+\delta_1(x))$. The set $Y_n \setminus Z_n$ is clearly nowhere dense and we take this set as our set N_n. The function δ_1 is a guage on Z_n and

this induces a denumerable partition $\{Z_{nk}\}$: we may see by the nature of this partition that if we define $c = \inf Z_{nk}$ and $d = \sup Z_{nk}$ then Y_n must be dense in the interval (c,d). Thus we define $Y_{nk} = Z_{nk}$ and this gives the partition (*) above, with the asserted properties.

Let N be the union of the sets N_n and let the sequence $\{X_n\}$ of the lemma be a relabelling of the doubly indexed sequence $\{Y_{nk}\}$. Certainly N is a first category set and so the lemma is proved if we are able to prove that the function f_r is increasing on each set Y_{nk}.

In fact taking any particular set Y_{nk} and setting $c = \inf Y_{nk}$ and $d = \sup Y_{nk}$ we will show that the inequality

$$\frac{f(y) - f(x)}{y - x} \geq r \qquad (**)$$

must hold everywhere for x, y in Y_{nk}. To see this we give an indirect proof. If (**) fails then define the function $g(x) = f(x) - rx$ and observe that there must be points x_1 and y_1 in (c,d) with $x_1 < y_1$ such that $g(x_1) > g(y_1)$. Define the point x_0 by writing

$$x_0 = \sup\{x \in [x_1, y_1] : g(x) \geq g(x_1)\}.$$

Since f is continous, g also is continuous and we must have $g(x_0) = g(x_1)$ and $x_1 \leq x_0 < y_1$. Set $h = y_1 - x_0$.

Using the continuity of g and the fact that Y_n is dense in the interval (c,d) we may find a point x in Y_n in such a way that

$$0 < x_0 - x < (1 - p)h$$

and

$$g(x) > g(x_0) - (s - r)(1 - p)h.$$

By the nature of the partition and the porosity requirements we must have

$$\lambda(S_x, x, x + h) < ph$$

so that there is a point z in the set S_x that lies in the interval

$$(x + (1 - p)h, x + h).$$

For this we will have

$$f(z) - f(x) > s(z - x)$$

and

$$z > x + (1 - p)h > x_0.$$

Thus

$$g(z) = f(z) - rz \geq f(x) + s(z - x) - rz$$
$$= g(x) + (s - r)(1 - p)h > g(x_0) = g(x_1).$$

But this contradicts the choice of the point x_0. From this contradiction we have the inequality (**) and the lemma now follows.

(52.2) COROLLARY. Let f be a continuous function, and let S be a local system so that, at each point x, if $S \in S(x)$ then S has either right hand or left hand porosity less than 1 at x. Then if

$$(S) - \underline{D} f(x) > r$$

at every point of a set of the second category, there is an interval (a, b) on which the function $f_r(x) = f(x) - rx$ is increasing.

PROOF. This follows directly from the lemma.

(52.3) Example. We may apply this result to obtain a relative monotonicity theorem for continuous functions with very weak positive density derivatives. Let f be continuous and suppose that, at every point x of a second category set X, one may find a measurable set E_x that has positive lower density on one side at least at x and so that
$$\frac{f(y) - f(x)}{y - x} > 0 \quad \text{for each } y \in E_x, \ y \neq x.$$
Then f is nondecreasing on some interval.

(52.4) Example. In contrast to example (52.3) Ornstein [191] constructs a continuous function f for which the set
$$S_x = \{ y : \frac{f(y) - f(x)}{y - x} \geq 0 \}$$
has positive <u>upper</u> density on one side at least at the point x, for every value of x and yet the function f is not nondecreasing on any interval.

The lemma above, for continuous functions, permits a number of refinements in situations in which more information about the function f is given. In these cases a similar relative monotonicity theorem is available but the exceptional set N of the lemma may be more precisely expressed.

(52.5) LEMMA. Let f be a continuous function that satisfies a Lipschitz condition
$$|f(x) - f(y)| \leq M |x - y|,$$
and let S be a local system so that, at each point x, if $S \in S(x)$ then S has right hand porosity less than 1 at x. Then if $(S)\text{-}\underline{D} f(x) > r$ at every point x of a set X then there is a denumerable partition
$$X = N \cup X_1 \cup X_2 \cup X_3 \cup \ldots$$
of the set X such that N is σ-porous and the function $f_r(x) = f(x) - rx$ is increasing on each set X_n.

PROOF. The proof is almost identical to the proof of the next lemma and we may omit it.

(52.6) LEMMA. Let f belong to $C(\psi)$, that is to say f is a continuous function that satisfies an inequality of the form
$$|f(x) - f(y)| \leq \psi(|x - y|) \qquad (|x - y| \leq 1)$$
where ψ is a continuous increasing function on $[0, +\infty)$ for which $\psi(0) = 0$ and $\psi_+'(0) = +\infty$. Let S be a local system that so that,, at every point x, if $S \in S(x)$ then S has right hand porosity less than 1 at x. Then if $(S)\text{-}\overline{D} f(x) < r$ at every point x of a set X then there is a denumerable partition

$$X = N \cup X_1 \cup X_2 \cup X_3 \cup \ldots$$

of the set X such that N is σ-(ψ)-porous and the function $f_r(x) = f(x) - rx$ is decreasing on each set X_n.

PROOF. As in the proof of (52.1) we may suppose that $(S)\text{-}\overline{D}f(x) < s < r$ at each $x \in X$, and we define the set

$$S_x = \{ y : y = x \text{ or } \frac{f(y) - f(x)}{y - x} < s < r \},$$

for each $x \in X$. Since each such set is in $S(x)$ we may suppose, as in the proof of lemma (52.1) that S_x has right porosity at x less than some $p < 1$. We define a guage δ on X in such a way that

$$\lambda(S_x, x, x+t) < pt \quad \text{for } 0 < t < \delta(x)$$

and use δ to induce a denumerable partition of the set X in the standard manner, which we shall denote as $\{Y_n\}$.

We further partition each set Y_n so that

$$Y_n = N_n \cup Y_{n1} \cup Y_{n2} \cup Y_{n3} \cup \ldots \tag{*}$$

where N_n has positive (ψ)-porosity at each of its points and the sets Y_{nk} are chosen in the following manner. We take N_n as the set of points in Y_n at which the left hand (ψ)-porosity is positive; then there is a guage δ defined on the set $Y_n \setminus N_n$ so that for every $x \in Y_n \setminus N_n$ and every $0 < t < \delta(x)$,

$$\psi(\lambda(Y_n, x_0 - t, x_0)) < \nu t$$

where ν is taken as follows: firstly we define numbers θ and τ from the interval $(0,1)$ in such a way that

$$\tau < (s - r)(1 - \theta) \text{ and } \theta > p(1 + \tau).$$

Since $s > r$ and $0 < p < 1$ these are simple linear inequalities that may be solved. Then we take

$$\nu < \frac{\tau}{1 + \tau}.$$

This guage δ induces a partition of the set $Y_n \setminus N_n$ and it is this partition that we label as $\{Y_{nk} : k=1,2,\ldots\}$.

Let N be the union of the sets N_n and we consider that the sequence $\{X_n\}$ of the lemma is a relabelling of the doubly indexed sequence $\{Y_{nk}\}$. Certainly N is σ-(ψ)-porous and so the lemma is proved if we are able to prove that the function f_r is decreasing on each set Y_{nk}.

In fact taking any particular set $Y = Y_{nk}$ and any pair of points x_0, x_1 in Y_{nk} with $x_1 < x_0$; we claim that the inequality

$$\frac{f(x_0) - f(x_1)}{x_0 - x_1} \leq s$$

must hold at these points. To verify this inequality then we define

$$x_2 = \sup \{ z \in (x_1, x_0) : \frac{f(z) - f(x_1)}{z - x_1} \leq s \}$$

and we prove that $x_2 = x_0$. As f is continuous this forces the required inequality

$$\frac{f(x_0) - f(x_1)}{x_0 - x_1} \leq s$$

and our claim is proved. We obtain this now by a contradiction; if contrary to this $h = x_0 - x_2 > 0$ then we consider the interval

$$(x_2 - \psi^{-1}(\tau h), x_2)$$

where τ is as chosen above and ψ^{-1} is the inverse function to ψ. We will take it that $\psi^{-1}(\tau h) < \tau h$ which requires only that τ has been chosen sufficiently small.

Using the porosity condition on Y_n and the inequality

$$\frac{\psi[\psi^{-1}(\tau h)]}{h + \psi^{-1}(\tau n)} > \frac{\tau h}{(1+\tau)h} > \nu$$

we see that there must be points of Y_n in the interval $(x_2 - \psi^{-1}(\tau h), x_2)$. Thus let x_3 be any point from Y_n in that interval.

Again the set S_{x_3} satisfies a porosity requirement in the interval (x_3, x_0): the interval $(x_0 - \theta h, x_0)$ has relative length

$$\frac{\theta h}{h + \psi^{-1}(\tau h)} > \frac{\theta}{1+\tau} > p$$

so that there must be a point x_4 in S_{x_3} from the interval $(x_0 - \theta h, x_0)$. We will obtain our contradiction by proving that

$$\frac{f(x_4) - f(x_1)}{x_4 - x_1} \leq s.$$

Since x_4 is evidently greater than x_2 this will contradict the definition of x_2 and our desired inequality will have been proved.

Putting our various computations together and using the ψ - inequality of the lemma, we obtain the inequalities

$$f(x_4) - f(x_1) = [f(x_4) - f(x_3)] + [f(x_3) - f(x_2)] + [f(x_2) - f(x_1)] \leq$$
$$r(x_4 - x_3) + \psi(|x_3 - x_2|) + s(x_2 - x_1).$$

This leads to the inequalities

$$f(x_4) - f(x_1) \leq r(x_4 - x_3) + \tau h + s(x_2 - x_1)$$
$$\leq r(x_4 - x_2) + \tau h + s(x_2 - x_1)$$
$$\leq r(x_4 - x_2) + (s-r)(1-\theta)h + s(x_2 - x_1)$$
$$\leq r(x_4 - x_2) + (s-r)(x_4 - x_2) + s(x_2 - x_1)$$
$$\leq s(x_4 - x_1)$$

as required. From this the lemma now follows.

We give a definition now for an intersection condition that will supply the next relative monotonicity result. Note that the one-sided approximate systems have this intersection property.

(52.7) DEFINITION. A local system S will be said to have the intersection property $[Z_\lambda]$ for a positive number λ provided that for any choice

$$\{S_x : x \in X\}$$

from S for a set X there is a guage δ on X so that whenever $x, y \in X$ with

$$0 < y - x < \min\{\delta(x), \delta(y)\}$$

the sets S_x and S_y intersect as follows: one at least of the two intersections
$$S_x \cap S_y \cap [x - \lambda(y - x), x] \quad \text{or} \quad S_x \cap S_y \cap [y, y + \lambda(y - x)]$$
is nonempty.

(52.8) LEMMA. Let S be a local system that is filtering and has the intersection property $[Z_\lambda]$ for a positive number λ. If a function f satisfies everywhere on a set X the inequalities
$$m < (S)\text{-}\underline{D} f(x) \leq (S)\text{-}\overline{D} f(x) < M$$
then there is a denumerable partition $\{X_n\}$ of the set X such that, for any pair of points x and y in a set $X \cap \overline{X}_n$, the inequality
$$m - \lambda(M - m) < \frac{f(y) - f(x)}{y - x} < M + \lambda(M - m)$$
must hold.

PROOF. Define the sets
$$S_x = \{ y : y = x \text{ or } m < \frac{f(y) - f(x)}{y - x} < M \}$$
for each $x \in X$ and note that each set S_x must belong to the family $S(x)$. We use the intersection condition $[Z_\lambda]$ to induce a denumerable partition $\{X_n\}$ so that
$$S_x \cap S_y \cap I \neq \emptyset$$
whenever $x < y$ belongs to a set X_{skn} and I is one of the intervals $[y, y + \lambda(y-x)]$ or $[x - \lambda(y-x), x]$.

Take a point z in this intersection. We show the arithmetic just for the case
$$y < z < y + \lambda(y - x)$$
but the other cases are similar. Since
$$\frac{f(z) - f(x)}{z - x} > m \quad \text{and} \quad \frac{f(z) - f(y)}{z - y} < M$$
we have
$$f(y) - f(x) = f(z) - f(x) - [f(z) - f(y)]$$
$$> m(z - x) - M(z - y)$$
$$= m(y - x) - (M - m)(z - y) = [m - (M - m)\lambda](y - x).$$
Similarly
$$f(y) - f(x) = f(z) - f(x) - [f(z) - f(y)]$$
$$< M(z - x) - m(z - y)$$
$$= M(y - x) + (M - m)(z - y) = [M + (M - m)\lambda](y - x)$$
as required. The extension to the closure of these sets X_n in X is routine.

From this theorem we may immediately deduce the following corollaries.

(52.9) COROLLARY. Let S be a local system that is filtering and has the intersection property $[Z_\lambda]$ for a positive number λ. Then, if a function f satisfies everywhere on a set X the inequalities
$$-\infty < (S)\text{-}\underline{D}f(x) \leq (S)\text{-}\overline{D}f(x) < +\infty,$$
there is a denumerable partition $\{X_n\}$ of the set X such that f satisfies a Lipschitz condition on each $X \cap \overline{X}_n$.

(52.10) COROLLARY. Let S be a local system that is filtering and has the intersection property $[Z_\lambda]$ for every positive number λ. Then, if a function f satisfies everywhere on a set X the inequalities
$$r < (S)\text{-}\underline{D}f(x) \leq (S)\text{-}\overline{D}f(x) < +\infty,$$
there is a denumerable partition $\{X_n\}$ of the set X such that for any pair of points x and y in a set $X \cap \overline{X}_n$ the inequality
$$\frac{f(y) - f(x)}{y - x} > r$$
must hold. A similar assertion holds for the upper derivates with the inequalitites reversed.

PROOF. Let us define the set of points X_{sk}, for $s > r$, by writing
$$X_{sk} = \{ x : s < (S)\text{-}\underline{D}f(x) \leq (S)\text{-}\overline{D}f(x) < k \}.$$
If we are able to produce a denumerable partition of each such set X_{sk} for which the required inequality holds then the result of the lemma itself must follow too. Define the number λ so that
$$s - \lambda(k - s) > r$$
and apply the lemma.

(52.11) Example. We may apply the results of this section to the approximate Dini derivatives and obtain a number of well-known properties of these derivates. For instance we know that the property $[Z_\lambda]$ is available for all $\lambda > 0$ and so a function f that has, at every point x in a set X, a finite approximate Dini derivatives from one side at least is Lipschitz on a sequence of sets, closed in X and covering X. In particular if f is continuous then f is ACG on X (cf. Saks [209, Theorem 10.14, p.239]). This same observation holds for the preponderant one-sided derivates as well since an intersection condition $[Z_\lambda]$ is available for some $\lambda > 0$.

§53. Relative monotonicity (cont.).

We continue the concerns of the preceding sections by obtaining two more specialized monotonicity theorems. These concern the approximate Dini derivatives and the negligent Dini derivatives. For the first of these results we use a weaker system than that for the approximate Dini derivatives, requiring only positive lower density on one side.

(53.1) LEMMA. Let S be a local system such that, at each point x, if $S \in S(x)$ then S has interior, lower density on one side at least that is positive, and suppose that a function f satisfies $(S)\text{-}\underline{D} f(x) > r$ at every point of a set X. Then X may be expressed as the union
$$X = N \cup X_1 \cup X_2 \cup X_3 \ldots$$
where N has measure zero and $f_r(x) = f(x) - rx$ is nondecreasing on each X_i.

PROOF. The proof is omitted, but can be constructed in a manner similar to that of other results of this type. See also the proof of (62.3).

(53.2) LEMMA. Let **N** be a σ-ideal of sets that contains no interval. Suppose that a function f satisfies $\underline{D}_N{}^+ f(x) > r$ at each point of a set X. Then X may be expressed as the union
$$X = N \cup X_1 \cup X_2 \cup X_3 \ldots$$
where $N \in \mathbf{N}$ and $f_r(x) = f(x) - rx$ is nondecreasing on each X_i.

PROOF. The proof is again omitted.

(53.3) Example. We sketch in this example a further relative monotonicity result due to F.K.Liu [151]. This assertion arises not from information about the derivates of the function but from the variation of the function. Let f be a continuous function that is VBG* on a Borel set E. Then there is a denumerable partition of the set E,
$$E = N \cup E_1 \cup E_2 \cup E_3 \cup \ldots$$
so that each set $N, E_1, E_2, \ldots$ is a Borel set, $\mu_f(N) = 0$, and f is strictly monotonic on each set E_i.

To prove this one writes $E = A^+ \cup A^- \cup N_1 \cup N_2$ where
$$N_1 = \{ x : f'(x) = 0 \},$$
$$N_2 = \{ x : f'(x) \text{ does not exist, finitely or infinitely} \}$$
$$A^+ = \{ x : f'(x) > 0 \}$$
$$A^- = \{ x : f'(x) < 0 \}$$
and then set $N = N_1 \cup N_2$. By (42.4) and (44.2), $\mu_f(N) = 0$. On the set A^+ (and similarly on A^-), one defines a guage δ so that
$$f(y') < f(x) < f(y'') \quad \text{for } x - \delta(x) < y' < x < y'' < x + \delta(x).$$
The guage partition of A^+ and A^- then supply the sets E_n.

§54. Global monotonicity theorems.

We turn now to a consideration of several classes of global monotonicity results, that is to say theorems whose conclusion is that some function is everywhere increasing or nondecreasing. The method that we choose here is to present assertions that are quite general, and easy to prove, but which contain a general expression of a great many known theorems.

The technical content in each theorem is contained basically in the observation that, in order to prove that a function f is monotonic nondecreasing, one may show that the interval function
$$I \to \Delta f^-(I) = \max\{-\Delta f(I), 0\}$$
vanishes identically. Frequently this may be obtained by showing that the variation $V(\Delta f^-, S)$ vanishes for some local system S. Since there is a rather immediate connection between derivation statements and statements involving the variation, this allows an immediate connection between derivation properties of a function and its monotonicity properties.

The definition we need is related directly to the estimates that were developed in section 38 for subadditive interval functions.

(54.1) DEFINITION. Let T be a collection of nonnegative subadditive interval functions and let S be local system. We say that S has a __monotonicity property__ relative to the class T if for every interval I and every $\psi \in T$ the inequality
$$\psi(I) \leq V_I(\psi, S)$$
holds.

With this definition our monotonicity results mostly follow from this elementary lemma (cf. also (42.1)).

(54.2) LEMMA. Let ψ be a nonnegative interval function, and let S be a local system. Suppose that at every point x in a set X, $(S)-\overline{D}\psi(x) < \alpha$ then
$$V_I(\psi, S; X) < \alpha |X \cap I|$$
for every interval I.

PROOF. Let $\varepsilon > 0$ and choose an open set G so that G contains X and
$$|X \cap I| > |G \cap I| - \varepsilon/\alpha .$$
Define the collection C of intervals J for which
$$J \subset G \quad \text{and} \quad \psi(J) < \alpha |J| .$$
Because of the inequality for the derivate in the assertion of the lemma this collection must be an S-cover of the set X. But evidently then
$$V_I(\psi, S; X) \leq \text{Var}_I(\psi, C) < \alpha |I \cap G| < \alpha |I \cap X| + \varepsilon .$$
Since $\varepsilon > 0$ is arbitrary the lemma is proved.

From this lemma one easily derives the following corollaries, using the fact that the variation behaves as an outer measure.

(54.3) COROLLARY. Let ψ be a nonnegative interval function, and let S be a local system. Suppose that, at every point x in a set X, $(S)-\overline{D}\psi(x) = 0$. Then $V(\psi, S; X) = 0$.

(54.4) COROLLARY. Let ψ be a nonnegative interval function, and let S be a local system. Suppose that, at every point x in a set X, $(S)-\overline{D}\psi(x) < +\infty$ and that X has measure zero. Then $V(\psi, S; X) = 0$.

We may now state and prove the basic results of this section. Note that these four theorems are variants on the same theme, but with the conclusions expressed somewhat differently in each case. The first is a direct monotonicity theorem; the second gives an estimate on the "maximum" decline of the function, that is to say an estimate on how much the function fails to be monotonic. The third and four theorems below are parallel results for constancy theorems; the former gives conditions under which a function may be deduced to be constant, and the latter an estimate on the oscillation of the function under similar hypotheses. The proofs are indicated after the assertions of the four theorems.

(54.5) THEOREM. Let T be a collection of nonnegative subadditive interval functions, and let S be a local system that has the monotonicity property relative to the class T. Let f be a real function with the following three properties:
(i) $\Delta f^- \in T$,
(ii) $(S)-\underline{D} f(x) \geq 0$ almost everywhere,
(iii) $(S)-\underline{D} f(x) > -\infty$ ν_f-almost everywhere,
where ν_f is the (outer) measure
$$E \to \nu_f(E) = V(\Delta f^-, S; E).$$
Then f is monotonic nondecreasing.

(54.6) THEOREM. Let T be a collection of nonnegative subadditive interval functions, and let S be a local system that has the monotonicity property relative to the class T. Let f be a real function with the following two properties:
(i) $\Delta f^- \in T$,
(ii) $(S)-\underline{D} f(x) \geq 0$ almost everywhere.
Then
$$\sup_I -\Delta(f(I)) \leq \nu_f\Big(\{ x : (S)-\underline{D} f(x) = -\infty \}\Big)$$
where ν_f is the (outer) measure
$$E \to \nu_f(E) = V(\Delta f^-, S; E).$$

(54.7) THEOREM. Let T be a collection of nonnegative subadditive interval functions, and let S be a local system that has the monotonicity property relative to the class T. Let f be a real function with the following three properties:
(i) $|\Delta f| \in T$,
(ii) $(S)-\overline{D} |\Delta f|(x) = 0$ almost everywhere,
(iii) $(S)-\overline{D} |\Delta f|(x) < +\infty$ μ_f-almost everywhere,
where μ_f is the (outer) measure
$$E \to \mu_f(E) = V(\Delta f, S; E).$$
Then f is constant.

(54.8) THEOREM. Let τ be a collection of nonnegative subadditive interval functions, and let S be a local system that has the monotonicity property relative to the class τ. Let f be a real function with the following two properties:
(i) $|\Delta f| \in \tau$,
(ii) $(S)\text{-}\overline{D}|\Delta f|(x) = 0$ almost everywhere,

Then
$$\sup_I |\Delta f(I)| \leq \mu_f\big(\{x: (S)\text{-}\overline{D}|\Delta f|(x) = +\infty\}\big),$$
where μ_f is the (outer) measure
$$E \to \mu_f(E) = V(\Delta f, S; E).$$

PROOF. Each of these is an application of the basic lemma (54.2). For example to prove the first of these theorems let us define the three sets
$$X_1 = \{x: (S)\text{-}\underline{D} f(x) \geq 0\}$$
$$X_2 = \{x: -\infty < (S)\text{-}\underline{D} f(x) < 0\}$$
$$X_3 = \{x: -\infty = (S)\text{-}\underline{D} f(x)\}$$
Note that $\mathbb{R} = X_1 \cup X_2 \cup X_3$ and that $|X_2| = 0$.

Define the interval function ψ as
$$\psi(I) = \Delta f^-(I) = \max\{-\Delta f(I), 0\}$$
and note that $\overline{D}\psi(x) = 0$ for every $x \in X_1$, and that $\overline{D}\psi(x) < +\infty$ for every $x \in X_2$. By (42.4) and (42.1) we must have
$$V(\psi, S; X_1) = V(\psi, S; X_2) = 0.$$
By hypothesis $V(\psi, S, X_3) = 0$ and so
$$V(\psi, S; \mathbb{R}) \leq V(\psi, S; X_1) + V(\psi, S; X_2) + V(\psi, S; X_3) = 0.$$
Thus the variation must vanish as required to prove the theorem. The other three theorems have almost identical proofs.

§55. Applications.

The general pattern of monotonicity theorem given in the preceding section may be applied to a great many theorems. In this section we shall review some of the applications. Note that in each case the application requires only two verification steps. The first is to ascertain that the function f (or rather the associated interval function $|\Delta f|$ or Δf^-) to which the theorem is applied is a member of some given class τ of functions, and the second is to verify that the system S has the monotonicity property of Definition (54.1). For the latter we may use the material in section 38. We shall in each case give only the briefest indication of how the verification step might proceed; most of the applications are immediate.

Let us begin with some monotonicity theorems for the ordinary derivation. Thus we are using the system S_0 that expresses ordinary limits. As this system has the property of (38.1) any class of subadditive interval functions τ may be used. Thus no restrictions need be placed on the function except to ensure the vanishing of the negative variation measure ν_f. For example the semicontinuity conditions that are used in (55.3) below assure that this measure vanishes at each point of the denumerable exceptional set.

(55.1) If at every point x the lower derivate $\underline{D} f(x) \geq 0$ then f is monotonic nondecreasing.

(55.2) If at almost every point x the lower derivate $\underline{D} f(x) \geq 0$, and everywhere $\underline{D} f(x) > -\infty$ then f is monotonic nondecreasing.

(55.3) Let N be a denumerable set and suppose that at each point $x \in N$ one has
$$\limsup\nolimits_{h \to 0+} f(x - h) \leq f(x) \leq \liminf\nolimits_{h \to 0+} f(x + h) .$$
If at almost every point x, $\underline{D} f(x) \geq 0$, and at every point x except possibly at the points of N, the lower derivate $\underline{D} f(x) > -\infty$ then f is monotonic nondecreasing.

(55.4) Let f be lower ACG*, i.e. the measure
$$E \to \nu_f(E) = V(\Delta f^-, S_0; E)$$
vanishes at every set of (Lebesgue) measure zero. If at almost every point x the lower derivate $\underline{D} f(x) \geq 0$ then f is monotonic nondecreasing.

(55.5) Let f be continuous and VBG* and let N be a set for which the image measure vanishes, i.e. $|f[N]| = 0$. If at almost every point x the lower derivate $\underline{D} f(x) \geq 0$, and at every point x, except possibly at the points in N, $\underline{D} f(x) > -\infty$, then f is monotonic nondecreasing.

Completely analogous properties are available for the bilateral approximate derivates since the system that describes such limits again permits the application of lemma (38.1). Monotonicity theorems of this kind have been given by Burkill [45, p.275], Sunouchi and Utagawa [224], Goffman and Neugebauer [102], Preiss [195], and O'Malley [180]. Let us list a few such results.

(55.6) If at every point x the lower approximate bilateral derivate $\underline{D}_{ap} f(x) \geq 0$ then f is monotonic nondecreasing.

(55.7) If at almost every point x the lower approximate bilateral derivate $\underline{D}_{ap} f(x) \geq 0$, and everywhere $\underline{D}_{ap} f(x) > -\infty$ then f is monotonic nondecreasing.

(55.8) Let N be a denumerable set and suppose that at each point $x \in N$ one has
$$\text{ap-}\limsup\nolimits_{h \to 0+} f(x - h) \leq f(x) \leq \text{ap-}\liminf\nolimits_{h \to 0+} f(x + h) .$$
If at almost every point x, $\underline{D}_{ap} f(x) \geq 0$, and at every point x except possibly at the points of N, the lower derivate $\underline{D}_{ap} f(x) > -\infty$ then f is monotonic nondecreasing.

For the onesided derivates the function f in the monotonicity theorems must be more severely restricted. A number of variants are known; for example see Saks [209, p.201-203], Gál [79, pp.309-310], Ważeweski [247], and Garg [83]. Recall that for the system S_0^+ that expresses right hand limits lemma (38.4) provides conditions on a class of interval functions T so that a monotonicity property holds. If we translate our general theorems for this particular system we obtain the following well-known results. Note that the semicontinuity conditions that are used on the function f are again what is needed to ensure that the measure

$$E \to \nu_f(E) = V(\Delta f^-, S_0^+; E)$$

that appears in this setting vanishes on the exceptional sets. These conditions are frequently called the "Zygmund conditions" (see Saks [209, p.203]). We state right hand versions only; we use both the systems S_0^+ and its dual $(S_0^+)^*$ so that although the general theorem requires lower derivates this gives a version for the lower and a version for the upper Dini derivatives.

(55.9) Suppose that the function f satisfies the following conditions:
(i) at each point x,
$$\liminf_{h \to 0+} f(x-h) \le f(x) ,$$
(ii) $\underline{D}^+ f(x) \ge 0$ a.e. , and
(iii) $\underline{D}^+ f(x) > -\infty$ everywhere.
Then f is monotonic nondecreasing.

(55.10) [Titchmarch and Aumann] Suppose that the function f satisfies the following conditions:
(i) at each point x,
$$\limsup_{h \to 0+} f(x-h) \le f(x) ,$$
(ii) $\overline{D}^+ f(x) \ge 0$ a.e. , and
(iii) $\overline{D}^+ f(x) > -\infty$ everywhere except possibly at the points x of a denumerable set, at each point of which the inequality
$$f(x) \le \limsup_{h \to 0+} f(x+h)$$
holds.
Then f is monotonic nondecreasing.

(55.11) [Ważewski] Suppose that the function f is continuous and that $|f[Q]| = 0$ where Q is the set
$$Q = \{ x : \overline{D}^+ f(x) < 0 \} .$$
Then f is monotonic nondecreasing.

Let us turn now to the analogous problems for the approximate Dini derivatives. These results were obtained by Burkill [45], Császár [54], Ornstein [191], and O'Malley [175]. Again we have lemma (38.6) that permits us to formulate the following monotonicity theorems.

(55.12) [Burkill] Suppose that the function f satisfies the following conditions:
(i) at each point x,
$$\text{ap-lim sup}_{h \to 0+} f(x - h) \leq f(x),$$
(ii) $\underline{D}_{ap}^+ f(x) \geq 0$ a.e., and
(iii) $\underline{D}_{ap}^+ f(x) > -\infty$ everywhere.
Then f is monotonic nondecreasing.

(55.13) [O'Malley] Suppose that the function f belongs to the first Baire class and satisfies the following conditions:
(i) at each point x,
$$\text{ap-lim sup}_{h \to 0+} f(x - h) \leq f(x),$$
(ii) $\overline{D}_{ap}^+ f(x) \geq 0$ a.e., and
(iii) $\overline{D}_{ap}^+ f(x) > -\infty$ everywhere except possibly at the points x of a denumerable set, at each point of which the inequality
$$f(x) \leq \text{ap-lim sup}_{h \to 0+} f(x + h)$$
holds.
Then f is monotonic nondecreasing.

Analogous statements again are available for the negligent Dini derivatives. These are consequences of lemma (38.5). We suppose for both of these results that **N** is a given σ-ideal of sets of real numbers, that does not contain an interval. Other variants may be given.

(55.14) Suppose that the function f satisfies the following conditions:
(i) at each point x,
$$\text{N-lim inf}_{h \to 0+} f(x - h) \leq f(x),$$
(ii) $\underline{D}_N^+ f(x) \geq 0$ a.e., and
(iii) $\underline{D}_N^+ f(x) > -\infty$ everywhere.
Then f is monotonic nondecreasing.

(55.15) Suppose that the function f satisfies the following conditions:
(i) at each point x,
$$\text{N-lim sup}_{h \to 0+} f(x - h) \leq f(x),$$
(ii) $\overline{D}_N^+ f(x) \geq 0$ a.e., and
(iii) $\overline{D}_N^+ f(x) > -\infty$ everywhere.
Then f is monotonic nondecreasing.

In addition the reader may wish to consult a number of topological versions of these monotonicity theorems given by Császár [54].

§56. The Goldowski-Tonelli theorem.

Each of the monotonicity theorems of the preceding section may be viewed as a generalization of the theorem of the elementary calculus that a function that posseses a positive derivative must be increasing. Those generalizations drop the differentiability assumption and use some generalized extreme derivate together with, perhaps, a regularity assumption. The theorem first proved by Goldowski [105] and Tonelli [243] is a generalization in a different spirit.

THEOREM. [Goldowski/Tonelli] Let f be continuous and have at every point a finite or infinite derivative. Then, if $f'(x) \geq 0$ at almost every point, f must be nondecreasing.

Here differentiability (in the extended sense) is retained but the requirement that the derivative be <u>everywhere</u> positive is relaxed. This theorem plays a key role in the study of the properties of derivatives and is well known, particularly so, since it appears in Saks [209, pp.206-207] with a complete proof.

This theorem, too, has been subjected to generalization. Tolstov [242] showed that the ordinary derivative could be replaced by the approximate derivative. Zahorski [265] relaxed the continuity assumption in the original theorem and prompted an investigation as to whether this relaxation could be used in the Tolstov version. This was completed by Bruckner [26] and Świątkowski [225]. See also the discussions in Leonard [152], O'Malley [175, pp.84-86], and Garg [95]. The theorem we now present includes most special cases of the theorem.

(56.1) THEOREM. Let S be a local system that is filtering and has an intersection condition of the form
$$S_x \cap S_y \cap [x,y] \neq \emptyset.$$
We suppose that whenever ψ is a continuous, nonnegative, subadditive interval function, and I is an interval then
$$\psi(I) \leq V_I(\psi, S).$$
Then if f is a Baire 1 function that satisfies the conditions (i), (ii), and (iii) below f must be nondecreasing:

(i) at every point x,
$$\liminf_{h \to 0+} f(x-h) \leq f(x) \leq \limsup_{h \to 0+} f(x+h),$$
(ii) $(S)\text{-}Df(x)$ exists finitely or infinitely at every point x with at most denumerably many exceptions, and

(iii) $(S)\text{-}Df(x) \geq 0$ almost everywhere.

PROOF. For continuous f the proof may be constructed in almost the same manner as the classical proof (see Saks [209, p.206]). The extension to functions satisfying the hypotheses of the theorem is then accomplished by using Bruckner's reduction theorem (see Bruckner [33, p.181] and Garg [95]).

CHAPTER SIX

RELATIONS AMONG DERIVATES

§57. Introduction.

Since the early years of this century there have accumulated many results that concern the relations that must hold among a variety of generalized derivates and derivatives. The earliest contributions were given by Levi [148], Sierpiński [212], Rosenthal [207], W.H.Young [258], G.C.Young [257], and Denjoy [59] in the first two decades of the century, but results of this type continue to appear even to this day.

The theorems that provide the focus for our study are mostly well known and stated below. The first of these is due to W.H.Young; it did not, however, attract much attention until its rediscovery by Neugebauer [171] in 1962. The version which we quote here uses the sharp derivates and was discovered by Jurek [133] in 1937.

THEOREM. [W.H.Young/B.Jurek] If f is a continous function then the relations
$$\overline{D} f(x) = \overline{D}^{\#} f(x) \quad \text{and} \quad \underline{D} f(x) = \underline{D}^{\#} f(x)$$
hold for the bilateral and sharp bilateral derivates everywhere except possibly on a a set of the first category.

The second of these relations that we quote is due to Levi [148] in 1906 who observed that for a continuous function f with one-sided derivatives everywhere the set of points where the derivatives differ, i.e. the set
$$\{ x : f_+'(x) \neq f_-'(x) \},$$
must be denumerable. (A similar result appears as a detail in an article of Hilbert [117] also from 1906.) There was some criticism of Levi's proof and correct versions were later given by Sierpiński [212] and Rosenthal [207], both in 1912. The final version, stated here for the Dini derivates, is from Mrs. Young and appeared in 1914.

THEOREM. [G.C.Young] For an arbitrary function f the relations
$$\overline{D}^+ f(x) \geq \underline{D}^- f(x) \quad \text{and} \quad \overline{D}^- f(x) \geq \underline{D}^+ f(x)$$
hold for the Dini derivates everywhere except possibly at a denumerable set.

These two theorems are elementary and have transparent proofs but they serve to reveal the intention of most of this chapter. One wishes to obtain relations of this type for a great many derivates; there is a large literature now devoted to the investigation of such relations. Both of these theorems and their many generalizations assume a particularly simple form in our language and suggest a standardized method of proof. After some modest analysis it can be seen that, in each of these theorems, one is required to show that a set of the form
$$X = \{ x : (S_1) - \overline{D} f(x) < (S_2) - \underline{D} f(x) \}$$

is small in some sense for certain choices of the systems S_1 and S_2. This may be proved by using the following well known device. For rational numbers r and s with $r < s$ write
$$X_{rs} = \{ x : (S_1) - \overline{D} f(x) < r < s < (S_2) - \underline{D} f(x) \}$$
so that X is expressible as a denumerable union of such sets. For each x in X_{rs} one then defines the sets
$$S^1{}_x = \{ y : y = x \text{ or } \frac{f(y) - f(x)}{y - x} < r \},$$
and
$$S^2{}_x = \{ y : y = x \text{ or } \frac{f(y) - f(x)}{y - x} > s \},$$
which evidently express choices from S_1 and S_2 respectively. From these choices it is often routine to show that each set X_{rs} belongs to some prescribed σ-ideal and the desired relation will be established.

We consider many such theorems in this chapter and each of them may be displayed and proved in exactly this elementary fashion.

A further relational theorem that will concern us in this chapter, but which is of a slightly different nature, is due to Khintchine [137]. Originally stated for exact approximate derivatives it was extended to the approximate Dini derivates by Mišík [166]; it is essentially this version that we quote here. Note that here the relation is not asserted to exist at all but an exceptional set, as in the two theorems above, but is required to hold at each point.

THEOREM. [Khintchine-Mišík] Let the function f be either monotonic or Lipshitz. Then at any point x_0 there must be the agreement
$$\overline{D} f(x_0) = \overline{D}_{ap} f(x_0) \quad \text{and} \quad \underline{D} f(x_0) = \underline{D}_{ap} f(x_0)$$
between the extreme derivates and the extreme approximate derivates.

This relation turns out to depend not on density considerations, as it might appear, but on porosity computations only. In section 65 we expore these ideas in some detail.

Thus in this chapter we shall be concerned with an abstract investigation of these three just mentioned theorems and a variety of related theorems, all of which have as their focus the assertion of a relation between two different processes of derivation. These concerns are continued in chapter six where a collection of results related to the Denjoy-Young-Saks theorem is presented.

§58. Elementary relations.

In this section we give the basic definitions and collect a number of relations that are more or less immediate. For example, the order relation on the class of local systems immediately imposes an order on the derivates and this appears stated in lemma (58.2). We begin by recalling the definition of the extreme derivates of a function taken with respect to a local system.

(58.1) DEFINITION. Let S be a local system and let f be an arbitrary function. Then the extreme S-derivates of f at a point x_0 are defined by setting $(S)\text{-}\overline{D} f(x_0)$ as
$$\inf\nolimits_{S \in S(x_0)} \sup\nolimits_{y \in S, y \neq x_0} \frac{f(y) - f(x_0)}{y - x_0},$$
and $(S)\text{-}\underline{D} f(x_0)$ as
$$\sup\nolimits_{S \in S(x_0)} \inf\nolimits_{y \in S, y \neq x_0} \frac{f(y) - f(x_0)}{y - x_0}.$$
An exact S-derivative of f at x_0, if it exists, is any number c (including $\pm\infty$) such that, for any neighbourhood U of c, the set of points
$$\left\{ y : y = x \text{ or } \frac{f(y) - f(x_0)}{y - x_0} \in U \right\}$$
belongs to $S(x_0)$. In this case we write $(S)\text{-}D f(x_0) = c$, with the warning that the number c need not be unique, nor have an immediate relation with the two extreme (S)-derivates. The set of all (S)-derivatives of a function f at a point x_0 will be denoted as $(S)\text{-}\Delta(f, x)$.

The lemmas which follow give an elementary development of these ideas. Note that many similar results for system limits were presented earlier and have similar proofs. We omit these obvious proofs.

(58.2) LEMMA. Let S_1 and S_2 be two local systems and let f be an arbitrary function. If $S_1 \ll S_2$ then, at any point x
$$(S_1)\text{-}\underline{D} f(x) \leq (S_2)\text{-}\underline{D} f(x) \quad \text{and} \quad (S_1)\text{-}\overline{D} f(x) \geq (S_2)\text{-}\overline{D} f(x).$$

(58.3) LEMMA. Let S be an local system of sets and let f be an arbitrary function. Then, at every point x, one has the relations
$$\underline{D} f(x) \leq (S)\text{-}\underline{D} f(x) \leq \overline{D} f(x), \quad \text{and} \quad \underline{D} f(x) \leq (S)\text{-}\overline{D} f(x) \leq \overline{D} f(x).$$

The feature that we have previously obvserved for limits, whereby the dual system serves to reverse the upper and lower limits, arises in the same way here for the two extreme derivates. This is a useful device in the theory in that any statement for an extreme derivate, upper or lower, may be converted immediately into an assertion for the opposite derivate, but taken with respect to the dual system.

(58.4) LEMMA. Let S be a local system of sets and let S^* denote its dual. Then, for any function f, one has the relations
$$(S)\text{-}\overline{D} f(x) = (S^*)\text{-}\underline{D} f(x), \text{ and } (S)\text{-}\underline{D} f(x) = (S^*)\text{-}\overline{D} f(x)$$
at every point x.

(58.5) **LEMMA.** Let S_1 and S_2 be local systems of sets and let f be an arbitrary function. Then, at every point x, one has the relations
$$(S_1 \vee S_2)\text{-}\overline{D} f(x) = \max\{(S_1)\text{-}\overline{D} f(x), (S_2)\text{-}\overline{D} f(x)\}$$
and
$$(S_1 \wedge S_2)\text{-}\overline{D} f(x) = \min\{(S_1)\text{-}\overline{D} f(x), (S_2)\text{-}\overline{D} f(x)\}$$
and similarly the relations
$$(S_1 \wedge S_2)\text{-}\underline{D} f(x) = \min\{(S_1)\text{-}\underline{D} f(x), (S_2)\text{-}\underline{D} f(x)\}$$
and
$$(S_1 \vee S_2)\text{-}\underline{D} f(x) = \max\{(S_1)\text{-}\underline{D} f(x), (S_2)\text{-}\underline{D} f(x)\}.$$

(58.6) Example. This last lemma expresses, somewhat abstractly, the same phenomenon that occurs in these easy relations that hold among the extreme bilateral derivates and the Dini derivates:
$$\overline{D} f(x) = \max\{\overline{D}^+ f(x), \overline{D}^- f(x)\},$$
and
$$\underline{D} f(x) = \min\{\underline{D}^+ f(x), \underline{D}^- f(x)\}.$$
We can consider that these arise because the systems S_0, S_0^+, and S_0^- are related by the assertion
$$S_0 = S_0^+ \vee S_0^-.$$

The upper derivates and the lower derivates in most general studies are related in an expected fashion: the upper derivate usually exceeds the lower derivate. In the present study it is by no means necessary that an upper derivate have a greater value than its corresponding lower derivate. The lemma gives an instance when this will happen. Recall that in chapter one, lemma (5.5), a similar result was discussed for the extreme limits; the idea here is identical.

(58.7) **LEMMA.** If S is filtering at a point x_0 then, for any function f, one has the relation
$$(S)\text{-}\overline{D} f(x_0) \geq (S)\text{-}\underline{D} f(x_0)$$
and $(S)\text{-}D f(x_0)$ if it exists must be unique. More generally these same assertions hold, provided only that at the point x_0, the system S has the property that whenever S_1 and S_2 are in $S(x_0)$ then $S_1 \cap S_2 \neq \{x_0\}$.

The order relation for the derivative and the exact derivative assumes the following somewhat peculiar form. Note that this next lemma, applied to the system S_∞, asserts merely the obvious fact that every derived number lies between the two extremes of the upper and lower derivates of a function.

(58.8) **LEMMA.** Let S be a local system and suppose that f is a function that has an exact (S)-derivative (S)-$D f(x_0)$ at a point x_0. Then one must have the relations
$$(S)\text{-}\overline{D} f(x_0) \leq (S)\text{-}D f(x_0) \leq (S)\text{-}\underline{D} f(x_0).$$
In particular if S is filtering at x_0 (or has the property expressed in lemma (58.7) above) then equality must hold.

The notion of derived number arises in a general way in this setting. The derived numbers, in the classical sense, of a function f at a point x are simply all the exact S_∞ derivatives of f at x. Generally, for a system S, the exact S^* derivatives play some kind of role as derived numbers. If S is filtering then the relation is quite close. The next two lemmas examine this situation.

(58.9) **LEMMA.** Let S be a local system and f an arbitrary function. Then, at any point x_0, the two numbers
$$(S)\text{-}\overline{D} f(x_0) \quad \text{and} \quad (S)\text{-}\underline{D} f(x_0)$$
are derived numbers of f at x_0. Moreover, if f has an S derivative (S)-$D f(x)$ at a point x, then that too is a derived number of f at x.

PROOF. Set $c = (S)\text{-}\overline{D} f(x)$ and for each n ($n = 1, 2, 3, \ldots$) consider the sets
$$A_n = \{ y : y = x \text{ or } \frac{f(y) - f(x)}{y - x} < c + \frac{1}{n} \} \cap (x - \frac{1}{n}, x + \frac{1}{n})$$
and
$$B_n = \{ y : y = x \text{ or } \frac{f(y) - f(x)}{y - x} < c - \frac{1}{n} \} \cap (x - \frac{1}{n}, x + \frac{1}{n}).$$
We know that $A_n \in S(x)$ but that B_n cannot belong, and that $A_n \supset B_n$.

Thus we may choose a point x_n in A_n but not in B_n with $x_n \neq x$. For this sequence it must be the case that
$$\frac{f(x_n) - f(x)}{x_n - x} \to c$$
and this shows that c must be a derived number of f. Similar arguments apply to the lower derivate and to the exact derivative in this sense.

(58.10) **LEMMA.** Let S be a local system that is filtering at a point x. Then if c is an extreme S-derivate of a function f at the point x, i.e.
$$(S)\text{-}\overline{D} f(x) = c \quad \text{or} \quad (S)\text{-}\underline{D} f(x) = c,$$
then c is an exact S^*-derivative of f at x, i.e.
$$(S^*)\text{-}D f(x) = c.$$
Indeed the upper derivate $(S)\text{-}\overline{D} f(x)$ is the greatest of such S^*-derivatives and the lower derivate is the least.

PROOF. Set $c = (S)\text{-}\overline{D} f(x)$ and for each n ($n = 1, 2, 3, \ldots$) consider the sets
$$A_n = \{ y : y = x \text{ or } \frac{f(y) - f(x)}{y - x} < c + \frac{1}{n} \}$$
and

$$B_n = \{ y : y = x \text{ or } \frac{f(y) - f(x)}{y - x} \leq c - \frac{1}{n} \}.$$

We know that $A_n \in S(x)$ but that B_n cannot belong, and that $A_n \supset B_n$.

If B_n does not belong to $S(x)$ then by the definition of the dual the set $(\mathbb{R} - B_n) \cup \{x\}$ must belong to the dual S^*. Since S is filtering the set
$$[A_n \cap (\mathbb{R} - B_n)] \cup \{x\}$$
must also belong to the dual S^*. But this simply says that the set
$$\{ y : y = x \text{ or } \left| \frac{f(y) - f(x)}{y - x} - c \right| < \frac{1}{n} \}$$
belongs to S^* for all n and consequently c is an exact S^* derivative of f at x.

§59. Beppo Levi theorem.

Recall the original theorem of Beppo Levi [148] from 1906.

THEOREM. [Levi] Let f be a continuous function that has one-sided derivatives everywhere on a set X. Then the set of points
$$\{ x \in X : f_+'(x) \neq f_-'(x) \}$$
at which the two derivatives differ must be denumerable.

After some analysis this theorem can be considered to have a elementary structure asserting an order relation between two extreme derivates, and not requiring any hypothesis on the function such as continuity or the existence of the one-sided derivatives. This analysis is due to G.C. Young who proves that, for any function f, the set of points
$$\{ x : \overline{D}^+ f(x) < \underline{D}^- f(x) \text{ or } \overline{D}^- f(x) < \underline{D}^+ f(x) \}$$
must be denumerable. This can, of course, be written positively as asserting that the set
$$\{ x : \overline{D}^+ f(x) \geq \underline{D}^- f(x) \text{ and } \overline{D}^- f(x) \geq \underline{D}^+ f(x) \}$$
encompasses all points with the exception only of a denumerable set.

Viewed abstractly this result assumes this form. There are two systems S_1 and S_2 given and one is required to assert that the set
$$\{ x : (S_1) - \overline{D} f(x) > (S_2) - \overline{D} f(x) \}$$
is small in some sense. Specifically with S_1 taken as a right-hand neighbourhood system and S_2 taken as the dual of the left-hand neighbourhood system this would give the G.C. Young result.

A number of classical results may be interpreted in this form, and we shall give a variety of theorems of this type. The first of these may be considered a direct generalization of the original theorem of Beppo Levi. Recall that, by an intersection condition on a set X for a system S, we mean that given any choice
$$\{ S_x : x \in X \}$$
from S (i.e. each $S_x \in S(x)$ for $x \in X$) there is a denumerable partition $\{ X_n : n = 1, 2, 3, \ldots \}$ of X with the property that for each pair $x, y \in X_n$ with $x < y$ the intersection

has some prescribed property. The simplest and most elementary results arise directly from an intersection condition of the form
$$S_x \cap S_y \cap [x,y] \neq \emptyset$$
and depend only on the monotonicity lemma (51.1) that is supplied by this condition: if, at every point x in a set X, the inequality
$$(S) - \underline{D} f(x) > r \qquad [\text{ respectively } (S) - \overline{D} f(x) < s]$$
holds, then there is a denumerable partition $\{X_n\}$ of the set X so that for any pair of points x and y from a set X_n one must have
$$\frac{f(x) - f(y)}{x - y} > r \qquad [\text{ resp. } \frac{f(x) - f(y)}{x - y} < s].$$

From this we obtain easily the following relation.

(59.1) THEOREM. Let S_1 and S_2 be two systems each of which satisfies an intersection condition of the form
$$S_x \cap S_y \cap [x,y] \neq \emptyset.$$
Then, for any function f, the set of points
$$\{ x : (S_1)-\underline{D} f(x) > (S_2)-\overline{D} f(x) \}$$
is denumerable.

PROOF. To show that the exceptional set of the theorem is denumerable it is enough, as usual, to show that for each rational number r the set of points
$$X_r = \{ x : (S_1)-\underline{D} f(x) > r > (S_2)-\overline{D} f(x) \}$$
is denumerable.
By two applications of the lemma above we obtain a denumerable partition $\{X_{rn}\}$ of the set X_r so that for any pair of points x and y from a set X_r the two inequalities
$$\frac{f(x) - f(y)}{x - y} < r < \frac{f(x) - f(y)}{x - y}$$
must hold. Since both of these inequalities cannot occur simultaneously we must conclude that each set X_{rn} can contain no more than a single point. From this it follows that the original set X_r is denumerable and the theorem is proved.

We obtain a number of corollaries from this result each asserting that some agreement between derivates or derivatives must occur nearly everywhere (i.e. with at most denumerably many exceptions).

(59.2) COROLLARY. Let S be a local system that satisfies an intersection condition as in the theorem. Then at every point x, with at most denumerably many exceptions, the order $(S)-\underline{D} f(x) \leq (S)-\overline{D} f(x)$ must hold.

PROOF. This follows from the theorem with $S = S_1 = S_2$.

(59.3) COROLLARY. Let S be as in the theorem. Then a function f has at every point x, with perhaps a denumerable number of exceptions, at most a single (S)-derivative. That is to say the set
$$(S) - \Delta(f, x)$$
is empty or a singleton nearly everywhere.

PROOF. This follows from the preceding corollary.

(59.4) COROLLARY. Let S_1 and S_2 be as in the theorem. Then if (S_1)-D $f(x) = g_1(x)$ and (S_2)-D $f(x) = g_2(x)$ everywhere on a set X the set of points
$$\{ x \in X : g_1(x) \neq g_2(x) \}$$
is denumerable.

PROOF. This follows from the preceding considerations.

(59.5) Example. This theorem and its corollaries apply to the right and left hand neighbourhood systems S_0^+ and S_0^- to give the classical theorems of Levi and G.C. Young. Numerous other systems can be used as well; one needs only a pair of systems, each of which permits an appropriate intersection property.

(59.6) Example. The density basis S_{ap} (sets having inner density 1) has this intersection condition. Thus one can conclude that, for any function f, the set of points
$$\{ x : \overline{D}_{ap} f(x) < \underline{D}^+ f(x) \}$$
must be denumerable. A variety of such conclusions is available.

This would even be the case for the preponderant derivate based on the density system that uses instead sets having inner, lower density exceeding 1/2 on both sides. However the onesided approximate derivates (approximate Dini derivates) based on a one sided density system (say right inner, lower density 1) do not have this property as they do not satisfy an intersection condition of this type. The first investigation of the approximate Dini derivates with a view to establishing a theorem of G.C. Young type was given by Ward [245]. We shall give an account of his ideas and results in section 62.

§60. Order relation for the negligent Dini derivates.

We continue the concerns of the preceding section by giving a further relational theorem that may be considered a generalization of the Levi theorem. Recall the system S_N that is generated by a σ-ideal of sets; this was discussed in detail in section 11. The special case with N taken as the σ-ideal of first category sets is called the "qualitative" system and the derivates relative to this system referred to as the qualitative derivates. We have defined right and left hand versions of these derivates in the qualitative sense or, more generally, relative to any σ-ideal N.

Recall that we write $\underline{D}^+_N f(x)$ as the supremum of all numbers M such that the set
$$(x, x+\delta_0) \setminus \{y : \frac{f(y)-f(x)}{y-x} < M\}$$
belongs to the σ-ideal for some positive number δ_0. This is the lower right N-Dini derivative; similar definitions supply the other three N-Dini derivatives. These derivates then are clearly system derivates relative to onesided versions of the system S_N.

For these derivates the Levi theorem takes the following form. Here the exceptional set need not be denumerable, but may differ from a denumerable set only by a set that belongs to the σ-ideal N itself.

(60.1) THEOREM. Let N be a σ-ideal of subsets of $\mathbb{R}$ that contains no interval. Then, for an arbitrary function f, each of the sets of points,
$$\{x : \underline{D}^+_N f(x) > \overline{D}^-_N f(x)\}$$
and
$$\{x : \underline{D}^-_N f(x) > \overline{D}^+_N f(x)\},$$
is the union of a denumerable set and a set that belongs to the σ-ideal N.

PROOF. As usual we define the set of points
$$X_{rs} = \{x : \underline{D}^+_N f(x) > r > s > \overline{D}^-_N f(x)\}$$
and apply the monotonicity lemma (53.2) in a manner similar to that given above.

(60.2) Example. The above relational theorem may be improved, as we shall see in section 63, by considering the set on which the derivates are finite. Otherwise the set may be an arbitrary member of the ideal N. For example take any set $N \in \mathbf{N}$ and define f as the characteristic function of the set N. Then for any $x \in N$ one has $\overline{D}_N^+ f(x) = -\infty$ and yet $\underline{D}_N^- f(x) = +\infty$.

§61. A further order relation.

In the same spirit as theorem (59.1) above, and with a similar proof, we may obtain a comparison between a weak density derivation (based on the system S_{wap} using sets having <u>exterior</u>, density 1) and any system that has the basic intersection condition. This theorem too can be considered as being of the same type as the G.C. Young theorem, in that an order is asserted to hold with the possible exception of some small set.

(61.1) THEOREM. Let S be a system that has an intersection condition of the form
$$S_x \cap S_y \cap [x, y] \neq \emptyset.$$
Then for any function f the sets of points
$$\{x : (S)\text{-}\overline{D} f(x) < \overline{D}_{wap} f(x)\}$$
and
$$\{x : (S)\text{-}\underline{D} f(x) > \underline{D}_{wap} f(x)\}$$
have measure zero.

PROOF. To show that the exceptional set of the theorem has measure zero it is enough to show that for each rational number r the set of points
$$X_r = \{ x : (S)\text{-}\underline{D} f(x) < r < (S_{wap})\text{-}\overline{D} f(x) \}$$
has measure zero. Because of the assumed intersection condition we may apply lemma (51.1) to obtain a denumerable partition $\{X_{rn}\}$ ($n = 1, 2, 3, \ldots$) of the set X_r such that for any pair of points x and y ($x < y$) in the same member X_{rn} one has the inequality
$$\frac{f(y) - f(x)}{y - x} < r.$$
At each point x of X_{rn} we have that $(S_{wap})\text{-}\overline{D} f(x) > r$ so that, in order for this set to allow the above inequality at each point $x \in X_{rn}$, there cannot be a point in X_{rn} at which the exterior density is equal to 1. By the Lebesgue density theorem this means that each set X_{rn} must then have measure zero as required. This completes the proof.

As a corollary we have a relation, at least for measurable functions, between certain system derivates and the usual approximate derivates of a function. This follows from the theorem on observing that for measurable functions the derivates relative to the system taking exterior density 1 are just the usual approximate extreme derivates.

(61.2) COROLLARY. Let S, f be as in the theorem but assume that f is measurable. Then, at almost every point x, one has the relations
$$(S)\text{-}\underline{D} f(x) \leq \underline{D}_{ap} f(x) \leq \overline{D}_{ap} f(x) \leq (S)\text{-}\overline{D} f(x).$$

(61.3) COROLLARY. Let S, f be as in the theorem and again assume that f is measurable. If the derivative $(S)\text{-}D f(x)$ exists everywhere on a set E then f is approximately differentiable almost everywhere on E and $f'_{ap}(x) = (S)\text{-}D f(x)$.

PROOF. This follows from the preceding corollary.

From the above considerations we see that a great many generalized derivations must agree almost everywhere with the approximate derivation process for measurable functions. For further, more delicate, observations of a similar nature see section 69.

§62. Ward's theorems.

A direct analogue of the theorem of Levi and G.C. Young is not available for the approximate Dini derivates. It has long been known that the set of points
$$\{ x : \overline{D}_{ap}^+ f(x) < \underline{D}_{ap}^- f(x) \}$$
at which the upper approximate Dini derivative on one side is less than the lower approximate Dini derivative on the other side need not be denumerable even for continuous functions. For measurable functions a trivial example shows this fact: let N be an uncountable measure zero set and let f be the characteristic function of N. Then, at every point x in N, the one sided approximate derivatives exist and violate the order that would be required for an analogue of the Levi theorem; indeed for each $x \in N$,

$$D^+_{ap} f(x) = -\infty \quad \text{and} \quad D^-_{ap} f(x) = +\infty.$$

Ward [245] and Zajíček [267] have investigated this situation and obtained a number of results. In this section we shall present what is known about these derivates as regards a relation of this type.

We first state Ward's examples.

(62.1) Example. [Ward] For any set E of measure zero there is an approximately continuous function f such that the set of points
$$\{ x : \overline{D}_{ap}^- f(x) < \underline{D}_{ap}^+ f(x) \}$$
includes the set E. (In particular the set of points at which this relation holds between the derivates need not be denumerable, nor first category, nor σ-porous.)

The construction is given in Ward [245, pp.295-297]. For the function f given there, in fact $D_{ap}^+ f(x) = +\infty$ and $D_{ap}^- f(x) = -\infty$ at every point x in E. This takes two pages of argument and one might ask whether a simpler construction may be found.

(62.2) Example. [Ward] Let E be any F_σ set of measure zero. Then there exists a continuous function f such that the set of points
$$\{ x : \overline{D}_{ap}^+ f(x) < \underline{D}_{ap}^- f(x) \}$$
includes the set E. In particular the set of points at which this relation holds between the derivates need not be denumerable, nor σ-porous, although it will be proved in (62.3) that it must be first category (as it is here). Zajíček [267] also gives an example but only to show that the set need not be countable. The construction appears in Ward [245, pp.297-298].

We may now state the positive results that were obtained by Ward. Although these results are interesting analogues of the G.C.Young theorem, the proper perspective is obtained by comparing them with the Denjoy-Khintchine theorem that appears in section 70. Indeed Ward simply modifed the Burkill, Haslam-Jones proof of that theorem to apply to nonmeasurable functions.

(62.3) THEOREM. For any function f the set of points
$$\{ x : \overline{D}_{ap}^+ f(x) < \underline{D}_{ap}^- f(x) \}$$
is of measure zero. If f is given to be continuous then this set is also of the first category.

PROOF. As usual we write for rational numbers r and s,
$$X_{rs} = \{ x : \overline{D}_{ap}^+ f(x) < r < s < \underline{D}_{ap}^- f(x) \}$$
and the theorem is proved by showing that each such set is measure zero, and also, should f be continuous, of first category. At each point $x \in X_{rs}$ construct the sets
$$S_x = \{ y : y = x \text{ or } \frac{f(y) - f(x)}{y - x} < r \} .$$
By the density requirements on S_x we may find a positive function δ on X_{rs} so that

$$|S_x \cap (x, x+t)|^i > \frac{3t}{4} \quad \text{for } 0 < t < \delta(x).$$

Let $\{X_{rsn}\}$ be the denumerable partition of the set X_{rs} that this function δ induces.

We show now that each set X_{rsn} has measure zero. Almost every point of X_{rsn} is both a point of density of X_{rsn} and a point of weak approximate continuity of f. Thus, if X_{rsn} has positive measure, we may find a point x_0 so that at x_0 the set X_{rsn} has exterior density 1 and such that there is a set A, that has also exterior density 1 at x_0, for which

$$\lim_{y \to x_0, y \in A} f(y) = f(x_0).$$

Let us choose the positive number δ_0 so that

$$|X_{rsn} \cap (x-t, x)|^e > \frac{3t}{4} \quad \text{for } 0 < t < \delta_0$$

and

$$|A \cap (x-t, x)|^e > \frac{3t}{4} \quad \text{for } 0 < t < \delta_0.$$

By the nature of the partition we may select a point x_1 so that $x_1 \in X_{rsn}$, so that $x_0 - \delta_0 < x_1 < x_0$, and so that

$$S_{x_1} \cap X_{rsn} \cap (x_1 + \frac{x_0 - x_1}{2}, x_0)$$

is nonempty. Take x_2 in this intersection and repeat the arguments inductively to obtain a sequence of points $\{x_m\}$ with $x_m < x_0$, $x_m \to x_0$, $x_m \in X_{rsn}$, and $x_m \in S_{x_{m-1}}$.

By a similar argument, repeated again for the set A which has the same density requirements, we take points z_m from the intersection

$$S_{x_m} \cap A \cap (x_m, x_0).$$

Here $f(z_m) \to f(x_0)$. Putting these facts together we find that

$$\frac{f(x_0) - f(x_1)}{x_0 - x_1} \leq r$$

for all such points x_1 in X_{rsn}. But this contradicts the fact that at the point x_0 the left hand approximate derivate is to exceed s. From this contradiction the first part of the theorem follows.

For the second part we assume that f is continuous and show that, morover, each set X_{rsn} is nowhere dense. This may be done directly, but as we will prove a more general version later (see (64.6)) the details may be omitted.

§63. Zajíček's relations.

A weak version of the G.C.Young theorem is available for the right and left approximate derivates. Let Y be the set of points where the approximate Dini derivatives are finite, i.e. Y is the set of points x at which

$$-\infty < \underline{D}_{ap}^+ f(x) \leq \overline{D}_{ap}^+ f(x) < +\infty$$

and

$$-\infty < \underline{D}_{ap}^+ f(x) \leq \overline{D}_{ap}^+ f(x) < +\infty.$$

Then, on that set, the appropriate relations hold with denumerably many exceptions, i.e. the set of points

$$X = \{ x \in Y : \overline{D}_{ap}^+ f(x) < \underline{D}_{ap}^- f(x) \}$$

must be denumerable.

This was first observed by Zajíček [267]. We can give, in fact, an abstract theorem of this type that depends only on the intersection condition $[Z_\lambda]$ given in section 52 to obtain a similar assertion. We obtain our result directly from the montonicity lemma (52.8). Note that the version here just takes advantage of the fact that the finiteness of the derivates and the intersection condition allows the estimates in lemma (52.8) to be made.

(63.1) THEOREM. Let S_1 and S_2 be local systems each of which is filtering and each of which has the intersection property $[Z_\lambda]$ for every positive number λ. Then for an arbitrary real function f the set of points at which all four of the derivates $(S_1)\text{-}\overline{D}f(x)$, $(S_1)\text{-}\underline{D}f(x)$, $(S_2)\text{-}\overline{D}f(x)$, and $(S_2)\text{-}\underline{D}f(x)$ are finite and yet
$$(S_1)\text{-}\overline{D}f(x) < (S_2)\text{-}\underline{D}f(x) \quad \text{or} \quad (S_1)\text{-}\underline{D}f(x) > (S_2)\text{-}\overline{D}f(x)$$
must be denumerable.

PROOF. We may consider the set of points x where
$$m < (S_1)\text{-}\underline{D}f(x) \leq (S_1)\text{-}\overline{D}f(x) < p < q$$
and
$$q < r < (S_2)\text{-}\underline{D}f(x) \leq (S_2)\text{-}\overline{D}f(x) < s.$$
By lemma (52.8) this set may be partitioned into a denumerable sequence of sets such that, for any pair of points x and y in the same member of the partition, both of the inequalities
$$\frac{f(y)-f(x)}{y-x} < q < \frac{f(y)-f(x)}{y-x}$$
hold. As this is impossible each such set can contain at most a single element and from this it is easy to see that the exceptional set of the theorem must be denumerable.

(63.2) Example. This result evidently applies to the approximate Dini derivatives, giving the result of Zajíček mentioned above. Thus we have that for an arbitrary real function f the set of points at which all four of the derivates $\overline{D}_{ap}^+ f(x)$, $\underline{D}_{ap}^+ f(x)$, $\overline{D}_{ap}^- f(x)$, and $\underline{D}_{ap}^- f(x)$ are finite and yet
$$\overline{D}_{ap}^+ f(x) < \underline{D}_{ap}^- f(x) \quad \text{or} \quad \underline{D}_{ap}^+ f(x) > \overline{D}_{ap}^- f(x)$$
must be denumerable.

(63.3) Example. Again the theorem applies to the negligient Dini derivatives. Using the notation we have previously established for these derivates we have that, for an arbitrary real function f, the set of points at which all four of the derivates $\overline{D}_N^+ f(x)$, $\underline{D}_N^+ f(x)$, $\overline{D}_N^- f(x)$, and $\underline{D}_N^- f(x)$ are finite and yet
$$\overline{D}_N^+ f(x) < \underline{D}_N^- f(x) \quad \text{or} \quad \underline{D}_N^+ f(x) > \overline{D}_N^- f(x)$$
must be denumerable.

§64. Theorem of W.H.Young.

The original theorem of W.H.Young from 1908 asserts that, for a continuous function f, the set of points at which there is a distinction between right and left as regards the derivates, i.e. the set of points
$$\{ x : \overline{D}^+ f(x) \neq \overline{D}^- f(x) \text{ or } \underline{D}^+ f(x) \neq \underline{D}^- f(x) \},$$
must be of the first category. This theorem is one of many that the Young family discovered for the left-right distinction of certain concepts. The theorem is rather more clearly revealed by expressing it not as a relation between right and left, but, as in the 1937 article of Jurek [133], by expressing it as asserting a relation between various derivates and the sharp derivates. (This same viewpoint appears in the later article of Bruckner and Goffman [36]). Thus we shall have that, for a continuous function f, the set of points at which
$$\{ x : \overline{D}^+ f(x) \neq \overline{D}^\# f(x) \text{ or } \underline{D}^+ f(x) \neq \underline{D}^\# f(x) \}$$
is of the first category. Similar results hold for the left derivates and for numerous other derivations.

Expressed in this way a number of general versions can be easily obtained. Our first generalization of Young's theorem merely uses an intersection condition and follows easily from the elementary monotonicity lemma (51.1); the second theorem is more subtle and arises from a porosity computation.

(64.1) THEOREM. Let S be a local system that has an intersection condition of the form
$$S_x \cap S_y \cap [x,y] \neq \emptyset .$$
Then, for any continuous function f, the set of points
$$\{ x : (S)\text{-}\underline{D} f(x) \neq \underline{D}^\# f(x) \text{ or } (S)\text{-}\overline{D} f(x) \neq \overline{D}^\# f(x) \}$$
is of the first category.

PROOF. To show that the exceptional set of the theorem is first category it is enough to show that, for each rational number r, the set of points
$$X_r = \{ x : (S)\text{-}\underline{D} f(x) < r < \underline{D}^\# f(x) \}$$
is first category. By an application of lemma (51.1), there is a denumerable partition $\{ X_{rn} \}$ of the set X_r so that, for any pair of points x and y from a set X_{rn},
$$\frac{f(y) - f(x)}{y - x} < r.$$
If any one of these sets X_{rn} is dense in some interval (c,d) then, by the continuity of the function f, the inequality
$$\frac{f(y) - f(x)}{y - x} \leq r$$
must hold for every pair of points x and y in (c,d). This is however impossible because there are points in this interval at which the derivate $\underline{D}_\# f(x) > r$. From this contradiction it follows that each set X_{rn} is nowhere dense and so that the exceptional set of the theorem is first category, as required.

By the same proof we may remove the hypothesis that f be continous and replace it with some one-sided continuity or semi-continuity conditions. This was first observed by Jurek [133]. Note that a function which satisfies the hypotheses of the corollary must, however, be continuous except for a denumerable set.

(64.2) COROLLARY. The theorem remains true if, in place of the assumption that f is continuous, it is given that, at every point x, one at least of the following four conditions holds:
(i) $f(x + 0) = f(x)$,
(ii) $f(x - 0) = f(x)$,
(iii) $\liminf_{t \to 0+} f(x+t) \geq f(x) \geq \limsup_{t \to 0+} f(x-t)$
(iv) $\liminf_{t \to 0+} f(x-t) \geq f(x) \geq \limsup_{t \to 0+} f(x+t)$.

PROOF. In the proof of the theorem observe that the inequality
$$\frac{f(y) - f(x)}{y - x} \leq r,$$
for x and y in the set X_{rn}, may be extended to any interval in which that set is dense using the conditions (i) - (iv) rather than the continuity assumptions. Thus essentially the same proof works under these hypotheses.

For a further corollary of this type observe that these semicontinuity conditions may be replaced by assumptions on the Dini derivatives. This gives us a further version of this theorem, again with almost the same proof.

(64.3) COROLLARY. Let S be as in the theorem but assume that the function f satisfies, at each point x of a set X, one at least of the following four conditions:
(i) $-\infty < \underline{D}^+ f(x) \leq \overline{D}^+ f(x) < +\infty$.
(ii) $-\infty < \underline{D}^- f(x) \leq \overline{D}^- f(x) < +\infty$.
(iii) $-\infty < \underline{D} f(x)$.
(iv) $+\infty > \overline{D} f(x)$.

Then the set of points in X,
$$\{ x \in X : (S)\text{-}\underline{D} f(x) \neq \underline{D}^\# f(x) \text{ or } (S)\text{-}\overline{D} f(x) \neq \overline{D}^\# f(x) \}$$
is of the first category in $\mathbb{R}$.

PROOF. Firstly let us split the set X into four sets X_1, X_2, X_3, and X_4 depending on which of the four conditions (i) - (iv) holds. We illustrate the proof by selecting one of these sets, say X_4, and arguing just on this set; the remaining sets are similarly handled.

For each integer m define the sequence
$$X_{4m} = \{ x \in X_4 : m > \overline{D} f(x) \}$$
and choose a positive number $\delta(x)$ for each $x \in X_{4m}$ so that
$$\frac{f(y) - f(x)}{y - x} < m \quad \text{for} \quad |x - y| < \delta(x), y \neq x.$$
The function δ induces a denumerable partition $\{ X_{4mn} \}$ of the set X_{4m} in our usual

way.

Now, holding m and n fixed, set $Y = X_{*mn}$ and argue exactly as in the proof of the theorem but just on the set Y. We leave the details to the reader.

(64.4) Example. This theorem and its corollaries show that most familiar derivations will have a relationship of this type. Thus one can apply the theorem to the Dini derivatives, to the approximate derivates, the preponderant derivates, and the qualitative derivates among others.

(64.5) Example. In the theorem and its corollaries some assumption such as continuity or semicontinuity, or some restrictions on the values of the derivates are required. A simple example shows this. Let f be the characteristic function of the irrationals. Then one computes easily that, at every point x,
$$\overline{D}^{\#} f(x) = +\infty \quad \text{and} \quad \underline{D}^{\#} f(x) = -\infty$$
and yet, at every irrational point x,
$$\overline{D}^{+} f(x) = 0 \quad \text{and} \quad \underline{D}^{-} f(x) = 0.$$

The Young theorem has been shown to apply to the approximate Dini derivatives by Pu, Chen and Pu [202]; Zajíček [271] has shown that it may apply as well to much thinner density derivatives. Neither of these results is contained in the general theorem (64.1) and so it is natural to ask for a more delicate version of that theorem. Perhaps the correct generalization of the theorem takes as a measure of the thickness required a very weak porosity computation, which then will include both the results of Pu et al. and the result of Zajíček. This has appeared in Thomson [236]; we repeat the result here with its proof.

(64.6) THEOREM. Let S be a local system of sets that has the property that, for each x and for each set $S \in S(x)$, S has porosity at x, on one side at least, less than 1. Then, for any continuous function f, the set of points
$$\{ x : (S)\text{-}\underline{D} f(x) \neq \underline{D}^{\#} f(x) \quad \text{or} \quad (S)\text{-}\overline{D} f(x) \neq \overline{D}^{\#} f(x) \}$$
is of the first category.

PROOF. The proof of the theorem may be reduced in the usual way to a consideration of the following: let X_{rs}, for fixed rational numbers r and s ($r < s$), denote the set of points x such that at each x
$$\underline{D}^{\#} f(x) < r$$
and yet each set
$$S_x = \{ y : y = x \quad \text{or} \quad \frac{f(y) - f(x)}{y - x} > s \}$$
has right porosity at x less than $p < 1$. We have merely to apply the monotonicity lemma (52.1) to obtain that each such set X_{rsp} is first category and the theorem evidently follows.

From this theorem we may obtain as a corollary the above mentioned result due to Zajíček [271]. We express this in the same language that Zajíček uses but note that, re-expressed in our compact language, it asserts a relation that holds for a density system that uses inner, upper density 1 (say S^{ap}): for a continuous function f the relations
$$(S^{ap})\text{-}\overline{D} f(x) = \underline{D}^{\#} f(x) \quad \text{and} \quad (S^{ap})\text{-}\underline{D} f(x) = \overline{D}^{\#} f(x)$$
must hold everywhere except possibly at the points of a first category set.

(64.7) COROLLARY. Let f be a continuous function. Then at every point x with the possible exception of a set of the first category there are measurable sets A_x and B_x each having upper density 1 at x and so that
$$\lim_{y \to x, y \in A_x} \frac{f(y) - f(x)}{y - x} = \overline{D}^{\#} f(x),$$
and
$$\lim_{y \to x, y \in B_x} \frac{f(y) - f(x)}{y - x} = \underline{D}^{\#} f(x).$$

PROOF. This follows directly from the theorem and the relation between porosity and density. The conversion here from system limits to actual path limits is then accomplished by the device of section 14.

Again this theorem may be extended to a slightly larger class of functions by making suitable modifications in the proof. We omit the details.

(64.8) COROLLARY. The theorem remains true if in place of the assumption that f is continuous it is given that, at every point x, one at least of the following four conditions holds:
(i) $f(x + 0) = f(x)$,
(ii) $f(x - 0) = f(x)$,
(iii) $\liminf_{t \to 0+} f(x + t) \geq f(x) \geq \limsup_{t \to 0+} f(x - t)$
(iv) $\liminf_{t \to 0+} f(x - t) \geq f(x) \geq \limsup_{t \to 0+} f(x + t)$.

(64.9) COROLLARY. Let S be as in the theorem but assume that the function f satisfies, at each point x of a set X, one at least of the following four conditions:
(i) $-\infty < \underline{D}^+ f(x) \leq \overline{D}^+ f(x) < +\infty$.
(ii) $-\infty < \underline{D}^- f(x) \leq \overline{D}^- f(x) < +\infty$.
(iii) $-\infty < \underline{D} f(x)$.
(iv) $+\infty > \overline{D} f(x)$.
Then the set of points in X,
$$\{ x \in X : (S)\text{-}\underline{D} f(x) \neq \underline{D}^{\#} f(x) \text{ or } (S)\text{-}\overline{D} f(x) \neq \overline{D}^{\#} f(x) \}$$
is of the first category in $\mathbb{R}$.

(64.10) Example. We illustrate the ideas of this section by indicating some possible applications. In many instances one can easily compute the extreme sharp derivates of a function. For example:

(i) if f is nowhere of bounded variation (i.e. has unbounded variation in each interval) then it may readily be shown that
$$\overline{D}^{\#} f(x) = +\infty \quad \text{and} \quad \underline{D}^{\#} f(x) = -\infty$$
at every point.

(ii) if f is strictly increasing and singular (i.e. $f'(x) = 0$ a.e.) then at every point x,
$$\overline{D}^{\#} f(x) = +\infty \quad \text{and} \quad \underline{D}^{\#} f(x) = 0.$$

(iii) if f has a dense set of discontinuities then at every point x
$$\overline{D}^{\#} f(x) = +\infty \quad \text{or} \quad \underline{D}^{\#} f(x) = -\infty .$$

These observations, together with the results of this section, allow one to make assertions that must hold residually for a wide variety of generalized derivatives. Thus, for instance, if f is a continuous function that is nowhere of bounded variation, then for most systems S, the equalities
$$(S)\text{-}\overline{D} f(x) = +\infty \quad \text{and} \quad (S)\text{-}\underline{D} f(x) = -\infty$$
hold residually.

§65. The porosity relations.

Khintchine [137] observed that for a monotonic function f the existence of the approximate derivative at a point entails that the function be in fact differentiable there. This property, in combination with other properties of the approximate derivative, can be used to obtain a great deal of information about the behaviour of an everywhere approximately differentiable function.

As this theorem of Khintchine plays such an important role in a number of investigations, it becomes natural to study it closer. Mišík [166] revealed that it should be considered as asserting an identity between the Dini derivates and the approximate Dini derivates for monotonic functions. Świątkowski [227] and Bruckner, O'Malley, and Thomson [43] (in different settings) consider the stronger proposition: what conditions are both necessary and sufficient in order that such a relation should hold. The correct formulation of the theorem turns out not to involve a density consideration, but may be expressed more economically by using the notion of porosity. We reproduce this theorem and some related results in this section.

The theorems that we wish to develop in this section all assert relations that must hold between various generalized derivates (based on systems satisfying some porosity requirement) and the ordinary derivates for certain classes of functions. The first result just gives that for monotonic functions or Lipschitz functions the "porosity zero" derivatives are precisely the ordinary derivatives. This is the most general version, for these classes of functions, of the theorems of Khintchine and Mišík cited above.

(65.1) THEOREM. Let f be a monotonic function or a function that satisfies a Lipshitz condition of the form
$$|f(x) - f(y)| \leq M |x - y| .$$

Suppose that at a point x_0 the system S is such that every set $S \in S(x_0)$ is nonporous at x_0. Then the relations
$$\overline{D} f(x_0) = (S)-\overline{D} f(x_0) \quad \text{and} \quad \underline{D} f(x_0) = (S)-\underline{D} f(x_0)$$
must hold.

PROOF. This is just a direct consequence of the porosity lemmas ($A_{5.1}$) and ($A_{5.3}$) given in the appendix. Note that one-sided versions hold too, although we shall not explicitly state them.

(65.2) Example. The theorems of Khintchine and Mišík follow directly from this theorem, since for the density system S_{ap} that expresses approximate derivation, any set $S \in S_{ap}(x_0)$ must have density 1 at x_0 and so, in particular, must be nonporous at x_0.

Thus, for any monotonic function f, if $f_{ap}'(x_0)$ exists at a point x_0, then f is necessarily differentiable at x_0. Similarly for such a function, and at every point, the four Dini derivatives and the four approximate Dini derivatives are identical.

A refinement of the theorem is also available for other porosity values.

(65.3) THEOREM. Let f be monotonic nondecreasing at a point x_0 and let p be a number, $0 \leq p < 1$. Suppose that at a point x_0 the system S has the property that every set $S \in S(x_0)$ has porosity on both sides smaller than p. Then one has the relations
$$(1-p)(S)-\underline{D} f(x_0) \leq \underline{D} f(x_0) \leq \overline{D} f(x_0) \leq \frac{1}{1-p}(S)-\overline{D} f(x_0).$$

PROOF. Suppose that $\underline{D} f(x_0) < r$. Then by the porosity lemma (Appendix ($A_{5.1}$)) the set of points
$$Y = \{ y : \frac{f(y)-f(x_0)}{y-x_0} \geq s > r \}$$
has porosity $1-r/s$ at least. If $(S)-\underline{D} f(x_0) > s$, then this set Y must be in $S(x_0)$, and so must have porosity less than p. This then requires that $1-r/s < p$, or equivalently
$$(1-p)s < r.$$
The first inequality in the theorem now follows; the second can be obtained in a similar fashion.

(65.4) COROLLARY. Let f be monotonic nondecreasing at a point x_0 and suppose that each set S in $S(x_0)$ has porosity less than 1 on both sides at x_0. Then, if $(S)-\underline{D} f(x_0) = +\infty$, it must be the case that $f'(x_0) = +\infty$. Similarly if $(S)-\overline{D} f(x_0) = 0$, it must be the case that $f'(x_0) = 0$.

PROOF. This follows directly from the inequalities in theorem (65.3).

We have stated the theorems above for monotonic functions. Similar assertions for Lipshitz functions may also be made.

(65.5) THEOREM. Let f satisfy a Lipshitz condition
$$|f(x) - f(y)| \leq M |x - y|$$
and suppose that every set S in $S(x_0)$ has porosity less than p at the point x_0. Then the relations
$$-pM + (1 - p) (S)-\underline{D} f(x_0) \leq \underline{D} f(x_0)$$
$$\leq \overline{D} f(x_0) \leq \frac{1}{1-p} (S)-\overline{D} f(x_0) + pM$$
must hold.

PROOF. This follows from the porosity computations $(A_{5.3})$ in the same way that theorem (65.3) was proved.

This theorem again extends to functions that satisfy Lipshitz conditions of a more general form.

(65.6) THEOREM. Let f satisfy a condition of the form
$$|f(x) - f(y)| \leq M |x-y|^\alpha \qquad (|x-y| \leq \delta_0)$$
for some numbers $0 < \alpha < 1$, $\delta_0 > 0$ and $M > 0$. Suppose that every set S in $S(x_0)$ has (x^α)-porosity less than t at the point x_0. Then the relations
$$-tM + (S) - \underline{D} f(x_0) \leq \underline{D} f(x_0)$$
$$\leq \overline{D} f(x_0) \leq (S) - \overline{D} f(x_0) + tM$$
must hold.

PROOF. Again this follows from $(A_{10.1})$.

Finally once again this theorem may be stated for a general class of functions $C(\psi)$ based on an appropriate function ψ that we use for a modulus of continuity. Here we will assume that the function ψ is continuous and is growing at 0 in a fashion similar to the function $\psi(t) = t^\alpha$ $(0 < \alpha < 1)$.

(65.7) THEOREM. Let the function f satisfy an inequality of the form
$$|f(x) - f(y)| \leq \psi(|x - y|) \qquad (|x - y| \leq \delta_0)$$
where ψ is continuous, increasing and has $\psi(0) = 0$ and $\psi'(0) = +\infty$. Suppose that every set S in $S(x_0)$ has (ψ)-porosity index less than t at the point x_0. Then the relations
$$-tM + (S) - \underline{D} f(x_0) \leq \underline{D} f(x_0) \leq \overline{D} f(x_0) \leq (S) - \overline{D} f(x_0) + tM$$
must hold.

PROOF. This, too, follows from the porosity computations in the appendix $((A_{10.1}))$.

§66. Evans-Humke theorem.

Evans and Humke [73] obtained a relation holding for the Dini derivatives of a monotonic function. Since monotonic functions have derivatives almost everywhere one needs a much smaller exceptional set in order to make a meaningful assertion. They give that, for a monotonic function f, the set of points
$$\left\{ x : \overline{D}^+ f(x) \neq \overline{D}^- f(x) \text{ or } \underline{D}^+ f(x) \neq \underline{D} f(x) \right\}$$
is σ-porous.

The theorem arises directly from an appropriate intersection condition, together with some basic material on porosity.

(66.1) THEOREM. Let f be monotonic and let S be a local system of sets that satisfies an intersection property of the form
$$S_x \cap S_y \cap [x,y] \neq \emptyset.$$
Then the set of points
$$\left\{ x : \overline{D} f(x) \neq (S)\text{-}\overline{D} f(x) \text{ or } \underline{D} f(x) \neq (S)\text{-}\underline{D} f(x) \right\}$$
is σ-porous.

PROOF. To show that the exceptional set of the theorem is σ-porous it is enough to show that, for each rational number r, the set of points
$$X_r = \{ x : (S)\text{-}\overline{D} f(x) < r < \overline{D} f(x) \}$$
is σ-porous. We define, for each $x \in X_r$, the set
$$S_x = \{ t : t = x \text{ or } \frac{f(t) - f(x)}{t - x} < r \}.$$
Applying the monotonicity lemma (51.1), we obtain a denumerable partition $\{X_{rn}\}$ of the set X_r so that, at each point x of X_{rn} and for every $y \in X_{rn}$, we have the inequality
$$\frac{f(y) - f(x)}{y - x} < r.$$
But at each point z of X_{rn} we also have that $\overline{D} f(z) > r$. By the porosity lemma (appendix $(A_{5.1})$), this means that the set X_{rn} must have positive porosity at each of its points. Consequently X_r is seen to be σ-porous and the theorem is proved.

(66.2) Example. For a Lipschitz function f the set of points
$$\{ x : \overline{D}^+ f(x) \neq \overline{D}^- f(x) \text{ or } \underline{D}^+ f(x) \neq \underline{D} f(x) \}$$
is σ-porous. This result is balanced by the following example of Evans and Humke [73]. They show that, for any set E that has measure zero and is first category, there is an absolutely continuous function f so that
$$E \subset \{ x : \overline{D}^+ f(x) \neq \overline{D}^- f(x) \}.$$

As before, the question arises as to whether a more refined verion of this theorem is available. Again we see that, just as in the refinement of the theorem of W.H. Young, what is needed is only a very thin porosity assumption on the family S. Stated as a relation it gives an assertion relating the ordinary derivates with derivates based on a system satisfying a weak porosity requirement.

(66.3) THEOREM. Let f be monotonic and let S be a local system of sets such that each set $S \in S(x)$ has porosity on the right or on the left at x less than 1. Then the set of points
$$\{ x : \overline{D} f(x) \neq (S)\text{-}\overline{D} f(x) \text{ or } \underline{D} f(x) \neq (S)\text{-}\underline{D} f(x) \}$$
is σ-porous.

PROOF. This follows easily from the monotonicity theorem (52.5).

From this we obtain easily a corollary of Zajíček [271] that asserts a relation between the density derivates based on a system of sets having upper density 1 and the ordinary derivates. This should be compared with (64.7) above.

(66.4) COROLLARY. Let f be monotonic or Lipschitz. Then at every point x, with the possible exception of a σ-porous set, there are measurable sets A_x and B_x each having uper density 1 at x such that
$$\lim_{y \to x, y \in A_x} \frac{f(y) - f(x)}{y - x} = \overline{D} f(x),$$
and
$$\lim_{y \to x, y \in B_x} \frac{f(y) - f(x)}{y - x} = \underline{D} f(x).$$

PROOF. This follows from the theorem in precisely the way that the corollary to theorem (64.7) was obtained.

This analysis will be carried even further in section 68 below. Note firstly that there is a strong relation between the W.H.Young theorem and the Evans-Humke theorem. For a continuous function f and any system S that is not too terribly thin (the sets have porosity less than 1 on at least one side) there is an agreement of the derivates
$$(S)\text{-}\overline{D} f(x) = \overline{D} f(x) \quad \text{and} \quad (S)\text{-}\underline{D} f(x) = \underline{D} f(x)$$
except on a set of the first category. If we strengthen the hypotheses so that f is even a Lipschitz function then the exceptional set may be improved from first category to σ-porous set. Thus the Evans-Humke theorem and the W.H.Young theorem appear to lie at two ends of some spectrum of theorems.

We can find this hierarchy between the two theorems by strengthening the hypotheses on f so that f is in a class $C(\psi)$ for some modulus of continuity ψ, that is so that f is a continuous function that satisfies a condition of the form
$$|f(x) - f(y)| \leq \psi(|x - y|) \qquad (|x - y| \leq 1).$$
For $\psi(t) = +\infty$, $\psi(0) = 0$ this inequality is vacuous and so f need only be continuous; for $\psi(t) = t$ this is the ordinary Lipschitz class. In general analogous theorems hold for the class of functions $C(\psi)$ but the exceptional set for this class of functions turns out to be a

σ-(ψ)-porous set. At one extreme this is to be interpreted as just a set of the first category, and for $\psi(t) = t$, at the other extreme, this is just a σ-porous set in the usual sense. These ideas are developed in section 68.

§67. Denjoy/Khintchine/Burkill/Haslam-Jones theorem.

The various density derivations have some strong relations. Denjoy [61, p209], Khintchine [137, p.212], Burkill and Haslam-Jones [46] have shown that the strongest of these derivations (the approximate derivation) is tightly related to one of the weakest (derivation taken relative to a system that uses merely sets having positive lower inner density). We state this theorem here. The assertion is not exactly as one usually finds it in the literature (cf. Saks [209, p.295]) but has been here rephrased so as to appear as a relation.

(67.1) THEOREM. Let f be a measurable function and let S be a simple system with the property that, for every point x and every $S \in S(x)$, the set S must have inner, lower density positive at x on one side at least. Then the relations
$$(S)\text{-}\overline{D}\, f(x) = \overline{D}_{ap}\, f(x) \quad \text{and} \quad (S)\text{-}\underline{D}\, f(x) = \underline{D}_{ap}\, f(x)$$
must hold almost everywhere.

PROOF. The proof is obtained directly from the monotonicity theorem (53.1). A complete proof appears too in Saks [209].

As corollaries we give the assertions that may be found in Saks.

(67.2) COROLLARY. If f is measurable on a set E and if, to each point x of E, there is given a set Q_x having positive inner lower density on one side at least at x and such that either
$$\overline{D}_{Q_x}\, f(x) < +\infty \quad \text{or} \quad \underline{D}_{Q_x}\, f(x) > -\infty ,$$
then f has a finite approximate derivative almost everywhere in E.

(67.3) COROLLARY. If f is measurable on a set E then at almost every point x of E there is a finite approximate derivative $f'_{ap}(x)$ or else
$$\overline{D}^+_{ap}\, f(x) = \overline{D}^-_{ap}\, f(x) = +\infty \quad \text{and} \quad \underline{D}^+_{ap}\, f(x) = \underline{D}^-_{ap}\, f(x) = -\infty .$$

This may also be obtained from (50.4).

§68. Generalized Young-Evans-Humke theorem.

The Evans-Humke theorem, as given in section 66, may be considered to belong in a hierarchy of theorems ranging from the theorem of W.H. Young, that for a continuous function f the set of points of right and left disagreement
$$\{ x : \overline{D}^+ f(x) \neq \overline{D}^- f(x) \text{ or } \underline{D}^+ f(x) \neq \underline{D}^- f(x) \}$$

is first category, through the various Lipschitz classes each time improving the exceptional set beyond merely first category to some σ-(ψ)-porous set.

This class of theorems permits another type of generalization. In place of discovering a comparison between the right and left ordinary Dini derivatives one can ask how far this extends to thinner derivates. We have seen in section 64 that the Young theorem extends to the approximate Dini derivatives, to extremely small density derivatives, and indeed even to very sparse porosity Dini derivatives. The extension to the full hierarchy of Lipshitz classes is given in Bruckner and Thomson [44]. We present this theorem here.

(68.1) THEOREM. Let f belong to C(ψ), that is to say f is a continuous function that satisfies an inequality of the form
$$|f(x) - f(y)| \leq \psi(|x - y|) \qquad (|x - y| \leq 1)$$
where ψ is a continuous increasing function on $[0, +\infty)$ for which $\psi(0) = 0$ and $\psi_+'(0) = +\infty$. Let S be a system such that at every point x and for every $S \in S(x)$ the set S has porosity less than 1 on one side at least at x.

Then at every point x, with the possible exception of a set that is σ-(ψ)-porous,
$$(S) - \overline{D} f(x) = \overline{D} f(x) \quad \text{and} \quad (S) - \underline{D} f(x) = \underline{D} f(x).$$

PROOF. The proof is obtained directly from the monotonicity theorem (52.6) and the porosity lemma ($A_{10.1}$) in our usual way.

As a corollary we may apply this theorem to certain density derivatives. We follow Zajíček [271] here in that we express the result as the existence of a type of path derivative along sets of upper density, but essentially this is nothing more than the observation that sets having positive lower (inner) density must have porosity less than 1.

(68.2) COROLLARY. Let f be a continuous function that satisfies an inequality of the form
$$|f(x) - f(y)| \leq \psi(|x - y|) \qquad (|x - y| \leq 1)$$
where ψ is as described in the theorem. Then at every point x, with the possible exception of a σ-(ψ)-porous set, there are measurable sets A_x and B_x each having upper density 1 at x and so that
$$\lim_{y \to x+, y \in A_x} \frac{f(y) - f(x)}{y - x} = \overline{D}^+ f(x) = \overline{D}^- f(x) = \overline{D} f(x)$$
and
$$\lim_{y \to x+, y \in B_x} \frac{f(y) - f(x)}{y - x} = \underline{D}^+ f(x) = \underline{D}^- f(x) = \underline{D} f(x).$$

PROOF. If at a point x such a set A_x could not be found then we can show that the set of points
$$\{ y : \frac{f(y)-f(x)}{y-x} < c < \overline{D} f(x) \}$$
must have some positive lower density, and so must have porosity less than 1. But we know from the theorem that the collection of such points has the asserted porosity requirement.

§69. Relations with the approximate derivative. In many instances one wishes to know whether a given generalized derivative must agree almost everywhere with the approximate derivative. We have already seen results in the preceding sections that can be used to answer such questions. In this section we shall briefly outline a few other ideas that could be pursued in connection with this kind of problem.

We repeat the content of corollary (63.3) here for reference.

(69.1) THEOREM. Let f be measurable and let S be a local system that has an intersection condition of the form
$$S_x \cap S_y \cap [x,y] \neq \emptyset.$$
If $(S)\text{-}Df$ exists everywhere on a set A then the approximate derivative of f exists a.e. on A and the equality
$$f_{ap}'(x) = (S)\text{-}Df(x)$$
holds almost everywhere on A.

One might wish to ask some weaker questions. There are two that are of some interest and for which partial answers are known. Suppose that f is a measurable function with a finite approximate derivative a.e. on a set A and that S is a local system. Under what conditions on S may we assert the following implications?

(a) If $(S)\text{-}Df(x)$ exists for a.e. x in A then
$$(S)\text{-}Df(x) = f_{ap}'(x) \qquad \text{for a.e. } x \in A.$$
(b) If $(S)\text{-}\overline{D}f(x) < +\infty$ for a.e. x in A then
$$(S)\text{-}\overline{D}f(x) = f_{ap}'(x) \qquad \text{for a.e. } x \in A.$$

Theorem (69.1) states that an intersection condition on the system S is enough to provide an answer to the question (a). Certainly any density requirements at all on the sets in the system would also supply the same result. It is of some interest to note that a porosity requirement would not be strong enough to answer either question even for continuous functions. We give this as an example.

(69.2) Example. We construct an example of a continuous function f and a system S such that
$$(S)\text{-}\underline{D}f(x) \geq 1$$
everywhere on a set X of positive measure, where each set $S \in S(x)$ is even nonporous on the right at x and yet the approximate derivative $f'_{ap}(x)$ exists and vanishes a.e. on X.

On the interval $[0,1]$ for any $1 > \varepsilon > 0$ we construct a sequence of continuous functions $\{f_n\}$ so that
$$0 \leq f_n \leq 1,$$
$$f_n\left(\frac{j}{4^n}\right) = 1, \qquad (j = 1, 2, \ldots 4^n - 1)$$
and
$$|\{x : f_n(x) \neq 0\}| < \frac{\varepsilon}{2^n}.$$

Then we write

$$f(x) = \sum_{n=1}^{\infty} \frac{1}{2^n} f_n(x).$$

Define $X = \{x : f(x) = 0\}$; clearly X is a closed nowhere dense set of positive measure and it is easy to verify that at each point $x \in X$, the set

$$S_x = \{y : y = x \text{ or } \frac{f(y) - f(x)}{y - x} \geq 1\}$$

is nonporous on the right at x.

In fact we can do considerably better than this example. We can arrange for the path derivative $(S)\text{-}Df(x)$ to exist, to be Baire 1, and to differ from the approximate derivative on a set of positive measure, even for a system that is nonporous on both sides. Thus we see that, in a question of this type, a porosity condition may not substitute for either an intersection condition or a density condition. A related example appears in Bruckner, Laczcovich, Petruska and Thomson [41].

From that paper, let us quote another result which gives an answer to question (a). The hypothesis here is that the system is of <u>congruent-type</u>, that is to say the system S is path generated (see section 14), and translation invariant (i.e. each $S(x)$ is the translation by x of the family $S(0)$).

(69.3) THEOREM. Let S be a congruent local system and let f be a measurable function that is approximately differentiable on a set A. If the derivative $(S)\text{-}Df(x)$ exists for a.e. x in A then

$$(S)\text{-}Df(x) = f_{ap}'(x) \quad \text{for a.e. } x \in A.$$

PROOF. This can be reduced to the following observations. If f is approximately differentiable on A then we can find a measurable subset A_1 of A with measure as close as we please to $|A|$ such that f has a derivative at each point of A_1 relative to A_1.

On the other hand, if $(S)\text{-}Df(x)$ exists a.e. on A there is a sequence $\{h_n\}$ converging to zero so that

$$(S)\text{-}Df(x) = \lim_{n \to \infty} \frac{f(x + h_n) - f(x)}{h_n}$$

for a.e. $x \in A$.

But a simple measure-theoretic argument will show that for almost every x in A_1 there are infinitely many indices n for which $x + h_n$ belongs to A_1. We can suppose that A_1 is bounded. Let $\varepsilon > 0$. Then there is a finite sequence of closed intervals $\{[a_i, b_i]\}$ so that

$$\left| A_1 \triangle \sum_{i=1}^{m} [a_i, b_i] \right| < \varepsilon.$$

Let us write $B_n = \{x \in A_1 : x + h_n \in A\}$. We show that $|B_n| \to |A_1|$ and the result evidently follows. If $x \in A_1$ and $0 < h_n < t$ but $x + h_n$ is not in A then either $x + h_n$ is in an interval $[b_i, b_i + t]$ or else $x + h_n$ is in the above small set difference. For sufficiently small t the set of such points has small measure and this proves our result.

Finally, putting these results together, we obtain the proof of the theorem. For details see Bruckner, Laczkovich, Petruska and Thomson [41].

Let us now turn to the question (b) above. This question was first addressed by Sindalovskiĭ in the setting of sequential derivatives. We use the notation
$$\overline{f'}_h(x) \quad \text{and} \quad \underline{f'}_h(x)$$
for the lim sup and lim inf as $n \to \infty$ of the expression
$$\frac{f(x + h_n) - f(x)}{h_n}$$
where $h = \{h_n\}$ denotes a fixed sequence of nonzero real numbers converging to zero. Sindalovskiĭ states the following result ([221, Lemma 3, pp.953-958]): Let $h = \{h_n\}$ be an arbitrary sequence of positive numbers converging to zero. If f is a measurable function on the interval $[0, 1]$, which has a finite approximate derivative everywhere on a set A, and which has
$$\overline{f'}_h(x) < +\infty$$
at every point x of the set A, then necessarily
$$f'_{ap}(x) = \overline{f'}_h(x)$$
at almost every point of A.

This is not the case. A sequence $h = \{h_n\}$ need not have this property. Bruckner, Laczkovich, Petruska, and Thomson [41] obtain a necessary and sufficient condition on a sequence h so that this would be true. We reproduce the condition here, but omit the rather lengthy proof.

(69.4) DEFINITION. A decreasing sequence $h = \{h_n\}$ of positive real numbers converging to zero will be said to satisfy the property (S) if the following is true. Whenever $P, P_1, P_2, P_3, P_4, \ldots$ are closed sets in the interval $[0, 1]$ such that
(a) every point $x \in P$ belongs to infinitely many of the sets P_i;
(b) $|P| > 0$;
(c) $P \cap [P_i + h_i] = 0$ for each index i,
then necessarily for every positive number C there are indices i and j with $i < j$, $h_i > Ch_j$ and
$$[P_i + h_i] \cap [P + h_j] \neq 0.$$

With this condition we may state an answer to question (b) in the setting of sequential derivatives.

(69.5) THEOREM. For a decreasing sequence $h = \{h_n\}$ of positive numbers converging to zero the following three assertions are equivalent.
(i) If f is a continuous function on the interval $[0, 1]$ such that everywhere on a measurable set A the approximate derivative $f'_{ap}(x)$ exists and $\overline{f'}_h(x) < +\infty$ then
$$f'_{ap}(x) = \overline{f'}_h(x) \quad \text{a.e. on } A.$$

(ii) If f is a measurable function on the interval [0,1] such that everywhere on a measurable set A the approximate derivative $f'_{ap}(x)$ exists and $\overline{f}'_h(x) < +\infty$ then
$$f'_{ap}(x) = \overline{f}'_h(x) \text{ a.e. on } A.$$
(iii) the sequence h has property (S) of definition (69.4).

Let us cite two further results from the article of Bruckner <u>et al.</u> [41] that are related to our concerns here.

(69.6) THEOREM. Let $h = \{h_n\}$ be a sequence of positive numbers such that
$$\limsup \frac{h_{n+1}}{h_n} < 1.$$
Then if f is a measurable function on the interval [0,1] such that
$$\overline{f}'_h(x) < +\infty$$
everywhere on a set A, and f is approximately differentiable a.e. on A, then a.e. on the set A,
$$\overline{f}'_h(x) = f'_{ap}(x).$$

(69.7) THEOREM. For every decreasing sequence of positive numbers $\{\alpha_n\}$ converging to zero there is a sequence $h = \{h_n\}$ of positive numbers converging to zero that does not have the property (S) and such that for every index k there is at least one index i for which
$$h_i \in (\alpha_k, \alpha_{k-1}).$$

CHAPTER SEVEN
THE DENJOY-YOUNG RELATIONS

§70. The Denjoy-Young-Saks theorem.

The most celebrated of the many theorems which assert relations among derivates is that known now as the Denjoy-Young-Saks theorem which gives a complete catalogue of the relations that must hold almost everywhere among the four Dini derivates. This theorem was first obtained for continuous functions (independently) by Denjoy [59] and G.C.Young in 1915. Mrs. Young delayed publication of her proof (for reasons that she explains in G.C.Young [257, p.361]) and then published a proof that the theorem would hold for measurable functions. The theorem was finally completed in 1924 by Saks [208] who extended it to apply to arbitrary functions. Since then the theorem has been much studied and there are a number of generalized versions known. In the literature the result is known generally under the various names, "the Denjoy relations", "the Denjoy-Young relations", or "the Denjoy-Young-Saks theorem".

The theorem is usually exhibited as asserting that, for an arbitrary function f, almost every point x must fall into one of the following four sets:

$$\{ x : f'(x) \text{ exists and is finite} \},$$
$$\{ x : \overline{D}^+ f(x) = \underline{D}^- f(x) \text{ are finite}, \underline{D}^+ f(x) = -\infty, \overline{D}^- f(x) = +\infty \},$$
$$\{ x : \underline{D}^+ f(x) = \overline{D}^- f(x) \text{ are finite}, \overline{D}^+ f(x) = +\infty, \underline{D}^- f(x) = -\infty \},$$

and

$$\{ x : \overline{D}^+ f(x) = \overline{D}^- f(x) = +\infty, \underline{D}^+ f(x) = \underline{D}^- f(x) = -\infty \}.$$

This analysis is not convenient for a method of proof nor for the task of formulating generalizations of the theorem. We may readily see that the following three theorems (together the obvious right/left and plus/minus versions) imply that the set of points that do not belong to one at least of the above four sets must be measure zero. The first of these theorems we have seen before in section 59.

[A] The set of points x at which $\underline{D}^+ f(x) > \overline{D}^- f(x)$ is denumerable.

[B] The set of points x at which
$$\overline{D}^+ f(x) = +\infty \quad \text{and} \quad \underline{D}^- f(x) \neq -\infty$$
has measure zero.

[C] The set of points x at which
$$+\infty > \overline{D}^+ f(x) > \underline{D}^- f(x) > -\infty$$
has measure zero.

These hold for an arbitrary function f and, by replacing $f(x)$ by $f(-x)$ and by $-f(x)$, the DYS theorem itself will follow.

In this chapter we present proofs of these basic theorems and discuss several variants of the Denjoy-Young relations that have been obtained by a number of investigators. For other references the reader may wish to consult the original articles of Denjoy, G.C. Young, and Saks cited above. See Saks [209, p.271] for a number of references to the early literature. In addition to these references one might consult Besicovitch [16], Blumberg [24], and Hanson [108]. Garg [80] gives a set of relations that holds with the exception of a set whose image has measure zero. Ravetz [203] explores the Hausdorff dimension of the exceptional set. The recent articles of Belna, Cargo, Evans and Humke [11], Preiss and Zajíček [198], and Zajíček [270], [271], [272] also continue these concerns.

§71. Proof of the DYS theorem.

We now prove the two theorems [B] and [C] stated in the introduction. The proof that we present is due basically to Hanson [108] and uses the Vitali theorem. For quite different proofs see Saks [209] or Jeffery [132].

(71.1) THEOREM. Let f be an arbitrary real function. Then the set of points at which either
$$\overline{D}^+ f(x) = +\infty \quad \text{and} \quad \underline{D}^- f(x) \neq -\infty,$$
or
$$\overline{D}^- f(x) = +\infty \quad \text{and} \quad \underline{D}^+ f(x) \neq -\infty,$$
or
$$\underline{D}^+ f(x) = -\infty \quad \text{and} \quad \overline{D}^- f(x) \neq +\infty,$$
or
$$\underline{D}^- f(x) = -\infty \quad \text{and} \quad \overline{D}^+ f(x) \neq -\infty,$$
is of measure zero.

PROOF. Let us show that the set of points X where
$$X = \{ x : \overline{D}^+ f(x) = +\infty \text{ and } \underline{D}^- f(x) > -\infty \}$$
has measure zero. The remaining parts of the theorem then will follow symmetrically. Thus for each integer m we will write
$$X_m = \{ x : \overline{D}^+ f(x) = +\infty \text{ and } \underline{D}^- f(x) > -m \}$$
and it will be sufficient to show that each member of the sequence $\{X_m\}$ has measure zero.

Define the collections of intervals
$$C_1 = \{ I : \Delta f(I) > -m |I| \}$$
and
$$C_k = \{ I : \Delta f(I) > k |I| \} \quad \text{for } k = 2, 3, 4, \ldots.$$
The collection C_1 is a S_0^- - cover of the set X_m while the collections C_k are S_∞^+ - covers of X_m. We use an intersection property of the system S_0^- (see example (15.8)) to find a denumerable partition $\{X_{mn}\}$ of the set X_m such that if x and y are

points with $y \in X_{mn}$, $x < y$, and x lies within the bounds of the set X_{mn} then the interval $[x, y]$ must belong to the cover C_1.

We use the cover C_k to estimate the Lebesgue measure of the set X_{mn} and we take advantage of the fact that such intervals $[x, y]$ as described above must belong to the cover C_1. Let $J = [a, b]$ be any interval with endpoints in X_{mn} and let A be the collection of all intervals from C_k whose left endpoint is in X_{mn} and which are subintervals of J. This collection A may be used to estimate the measure of the set $J \cap X_{mn}$.

Let π be an arbitrary finite set of nonoverlapping intervals chosen from A. By the nature of our construction the intervals in J that are complementary to π belong to the family C_1. Thus there is a collection π_1 from C_1 so that $\pi \cup \pi_1$ is a partition of the interval J.

Now we compute

$$(k+m) \sum_{I \in \pi} |I| < \sum_{I \in \pi} \Delta f(I) + m \left(b - a - \sum_{I \in \pi_1} |I| \right)$$

$$< \sum_{I \in \pi \cup \pi_1} \Delta f(I) + m(b-a)$$

$$< f(b) - f(a) + m(b-a).$$

As this gives an upper bound for the variation $\mathrm{Var}(\Delta x, A)$, we must have for the Lebesgue measure of the set $J \cap X_{mn}$ that

$$|J \cap X_{mn}|^e \leq \frac{f(b) - f(a) + m(b-a)}{k+m}.$$

This cannot hold for all integers k unless the set on the left of this inequality has measure zero. But this is the case for every interval with endpoints in X_{mn} so that X_{mn} itself has measure zero. Since the original set X is a denumerable union of measure zero sets it must have measure zero, as required to prove the theorem.

(71.2) THEOREM. For an arbitrary function f the set of points at which either

$$+\infty > \overline{D}^+ f(x) > \underline{D}^- f(x) > -\infty,$$

or

$$+\infty > \overline{D}^- f(x) > \underline{D}^+ f(x) > -\infty,$$

is of measure zero.

PROOF. We prove that the set of points X where

$$X = \{ x : +\infty > \overline{D}^+ f(x) > \underline{D}^- f(x) > -\infty \}$$

has measure zero. (Note that by the theorem of G.C. Young the set of points where

$$\overline{D}^+ f(x) < \underline{D}^- f(x)$$

is denumerable so that it would follow that in fact the set

$$+\infty > \overline{D}^+ f(x) \neq \underline{D}^- f(x) > -\infty,$$

X has measure zero.) The theorem will follow since the other set may be handled in a similar fashion.

Fix four rational numbers $p, q, r,$ and s and write
$$Y = \{ x : p > \overline{D}^+ f(x) > r > s > \underline{D}^- f(x) > q \}.$$
Evidently, if we are able to prove that any set Y of this form has measure zero, it will follow that the set X, which is a denumerable union of such sets, must also have measure zero.

Let us define four collections of intervals
$$C_1 = \{ I : \tfrac{\Delta f(I)}{|I|} < p \}$$
$$C_2 = \{ I : \tfrac{\Delta f(I)}{|I|} > q \}$$
$$C_3 = \{ I : \tfrac{\Delta f(I)}{|I|} > r \}$$
and
$$C_4 = \{ I : \tfrac{\Delta f(I)}{|I|} < s \}.$$
Note that C_1 and C_2 are respectively S_0^+- and S_0^--covers of the set Y. By using the special intersection properties of these two systems (see example (15.8)) we may form a denumerable partition $\{Y_n\}$ of the set Y that has the property that whenever $[x, y]$ is an interval within the bounds of a set Y_n then if $x \in Y_n$ the interval belongs to the collection C_1 while if $y \in Y_n$ then the interval belongs to the collection C_2.

Let us fix our attention on a particular set Y_n and let $J = [a, b]$ be an interval with endpoints in the set Y_n. We may estimate the Lebesgue measure of the set $Y_n \cap J$ by taking the collection A_3 of all intervals from C_3 whose left endpoint is in Y_n and which are subintervals of J; similarly we may estimate that same measure by taking the collection A_4 of all intervals from C_4 whose right endpoint is in Y_n and which are subintervals of J.

If π_3 is any finite collection of nonoverlapping intervals from A_3 then, by the nature of the construction, there must be a collection π_2 from C_2 so that together $\pi_2 \cup \pi_3$ forms a partition of the interval J. This gives the following elementary computation:

$$(r-q) \sum_{I \in \pi_3} |I| < \sum_{I \in \pi_3} \Delta f(I) - q\left(b - a - \sum_{I \in \pi_2} |I|\right)$$

$$< \sum_{I \in \pi_2 \cup \pi_3} \Delta f(I) - q(b-a) < f(b) - f(a) - q(b-a).$$

Similarly, for any π_4 that is a finite collection of nonoverlapping intervals from A_4, there must be a collection π_1 from C_1 so that together $\pi_4 \cap \pi_1$ is a partition of the interval J. Again this gives

$$(p-s) \sum_{I \in \pi_4} |I| < p\left(b - a - \sum_{I \in \pi_1} |I|\right) - \sum_{I \in \pi_4} \Delta f(I)$$

$$< p(b-a) - \sum_{I \in \pi_1 \cup \pi_4} \Delta f(I) < p(b-a) - f(b) + f(a).$$

These bounds for the variations $\mathrm{Var}(\Delta x, A_3)$ and $\mathrm{Var}(\Delta x, A_4)$ then give the following

bounds on the measure of the set $Y_n \cap J$,
$$(r - q) |Y_n \cap J|^e \leq f(b) - f(a) - q(b - a)$$
and
$$(p - s) |Y_n \cap J|^e \leq p(b - a) - f(b) + f(a)$$
After some arithmetic we obtain from these two inequalities
$$\frac{|Y_n \cap J|^e}{|J|} \leq \frac{p - q}{p - s + r - q} < 1.$$

As this inequality holds for every interval with endpoints in Y_n it follows that Y_n can have no points of density and so must, by the density theorem, have measure zero. From this the theorem now follows.

Note that an easy consequence of these relations is the well-known Lebesgue theorem asserting the a.e. differentiability of monotonic functions. This has also been proved in chapter four; note that both proofs use the Vitali theorem in some form.

(71.3) COROLLARY. Let f be monotonic or Lipshitz. Then f has a finite derivative almost everywhere.

PROOF. The fact that f is monotonic or Lipshitz excludes one or both of the values $+\infty$ and $-\infty$ from being obtained as derivate values. This then requires all four Dini derivates to agree except on a set of measure zero.

§72. The approximate version.

If, to the Denjoy-Young-Saks theorem, are added the relations available for the approximate Dini derivates, we obtain the analysis given in Burkill and Haslam-Jones [46] where, at least in the case of continuous functions, they attribute the ideas to Besicovitch.

Following the same pattern as given in the introduction we may assert that, for any measurable function f, almost every point x must fall into one of the following sets:
$$\{ x : f'_{ap}(x) \text{ exists and is finite} \}$$
and
$$\{ x : \overline{D}_{ap}{}^+ f(x) = \overline{D}_{ap}{}^- f(x) \text{ and } \underline{D}_{ap}{}^+ f(x) = \underline{D}_{ap}{}^- f(x) \}.$$

Moreover the points of the first of these sets may be exhibited as again lying, with the exception only of a set of measure zero, within one of the four sets:
$$\{ x : f'(x) \text{ exists and is finite} \},$$
$$\{ x : f'_{ap}(x) = \overline{D}^+ f(x) = \underline{D}^- f(x), \ \underline{D}^+ f(x) = -\infty, \ \overline{D}^- f(x) = +\infty \}$$
$$\{ x : f'_{ap}(x) = \overline{D}^- f(x) = \underline{D}^+ f(x), \ \underline{D}^- f(x) = -\infty, \ \overline{D}^+ f(x) = +\infty \}$$
and
$$\{ x : \overline{D}^+ f(x) = \overline{D}^- f(x), \ \underline{D}^+ f(x) = \underline{D}^- f(x) \}.$$

This version of the DYS theorem follows from the material in the preceding section together with one further relational theorem.

(72.1) THEOREM. Let f be a measurable function. Then the set of points at which either
$$\overline{D}_{ap} f(x) \neq \overline{D}^+ f(x) < +\infty, \qquad \overline{D}_{ap} f(x) \neq \overline{D}^- f(x) < +\infty,$$
$$\underline{D}_{ap} f(x) \neq \underline{D}^+ f(x) > -\infty \quad \text{or} \quad \underline{D}_{ap} f(x) \neq \underline{D}^- f(x) > -\infty,$$
is of measure zero.

PROOF. As usual we may obtain the proof of the theorem by showing that for any rational numbers p, r, and s the set of points.
$$X_{prs} = \{ y : p > \overline{D}^+ f(x) > r > s > \overline{D}_{ap}^+ f(x) \}$$
has measure zero. At each point x in X_{prs} consider the sets
$$S_x = \{ y : y = x \text{ or } \frac{f(y) - f(x)}{y - x} < p \}$$
and
$$T_x = \{ y : y = x \text{ or } \frac{f(y) - f(x)}{y - x} < s \}$$

There must be a guage δ on X_{prs} so that
$$S_x \supset (x, x + \delta(x))$$
and
$$|T_x \cap (x, x+t)|^i > (1 - \theta/2) t \quad \text{for} \quad 0 < t < \delta(x),$$
where $\theta = (p - r)/(p - s)$. Let $\{X_{prsn}\}$ be the denumerable partition of the set X_{prs} induced by the guage δ.

We show that each set X_{prsn} has measure zero and the theorem will follow. In order to obtain a contradiction let us suppose that some set X_{prsn} has positive measure and hence a point of density x_0. Choose a positive number $\delta_0 < \delta(x_0)$ and so that
$$|X_{prsn} \cap (x, x+t)|^e > (1 - \theta/2)t \quad \text{for} \quad 0 < t < \delta_0.$$
Since $\overline{D}^+ f(x_0) > r$ there must be a point x_1 in the interval $(x_0, x_0 + \delta_0)$ so that
$$\frac{f(x_1) - f(x_0)}{x_1 - x_0} > r.$$
By the density requirements on X_{prsn} and T_{x_0} there must be a further point x_2 in
$$X_{prsn} \cap T_{x_0} \cap (x_0 + \theta(x_1 - x_0), x_1)$$
and this requires that
$$f(x_2) - f(x_0) < s(x_2 - x_0).$$

Now we may compute
$$\begin{aligned}
f(x_1) - f(x_2) &= f(x_0) - f(x_2) + f(x_1) - f(x_0) \\
&> -s(x_2 - x_0) + r(x_1 - x_0) \\
&> s(x_1 - x_2) - s(x_1 - x_0) + r(x_1 - x_0) \\
&> (r - s)(x_1 - x_0) + s(x_1 - x_2) \\
&> \left[\frac{r - s}{1 - \theta} + s\right](x_1 - x_2) \\
&> p(x_1 - x_2)
\end{aligned}$$
But this latter contradicts the fact that x_2 is in X_{prsn} and from this contradiction we deduce our required result.

Zajíček [272] and Preiss and Zajíček [198] have investigated these relations and obtained narrower versions of the relations holding among the approximate Dini derivatives from the two following results.

(72.2) THEOREM. For an arbitrary function f the set of points x where
$$-\infty < \underline{D}_{ap}^+ f(x) \leq \overline{D}_{ap}^+ f(x) < \overline{D}_{ap}^- f(x),$$
or
$$-\infty < \underline{D}_{ap}^- f(x) \leq \overline{D}_{ap}^- f(x) < \overline{D}_{ap}^+ f(x)$$
is σ-porous.

PROOF. Let us define the set of points
$$X_{pqr} = \{ x : p < \underline{D}_{ap}^- f(x) \leq \overline{D}_{ap}^- f(x) < q < r < \overline{D}_{ap}^+ f(x) \}$$
for fixed rational numbers $p, q,$ and r. Clearly the theorem is proved if we are able to show that this set is σ-porous.

For each $x \in X_{pqm}$ define the sets
$$S_x = \{ y : p < \frac{f(y) - f(x)}{y - x} < q \}$$
and
$$T_x = \{ y : \frac{f(y) - f(x)}{y - x} > r \}.$$

Note that the set S_x must have (interior) density 1 on the left at x and that the set T_x must have (exterior) upper density positive on the right at x. For each integer m we may collect the set of points X_{pqrm} at which this latter density exceeds $1/m$, and define a guage δ on X_{pqrm} so that
$$|(x - t, x) \setminus S_x|^e < \frac{\rho t}{2m} \qquad \text{if } 0 < t < 2\delta(x),$$
where ρ denotes a number from the interval $(0, 1/2)$ chosen so that
$$\frac{2\rho}{1 - 2\rho} < \frac{r - q}{(r - p) + (q - p)}.$$

This guage δ induces a denumerable partition of the set X_{pqrm}, which partition we may denote by $\{ Y_n \}$.

We claim that the right porosity of each set $Y = Y_n$ at each of its points must be at least ρ. In order to obtain a contradiction let us suppose that there is a point x_0 in the set Y such that the right porosity of Y at x_0 is less than ρ, and hence there is a number δ_0 with
$$\lambda(Y, x_0, x_0 + t) < \rho t \qquad \text{if } 0 < t < \delta_0.$$
Choose a number $h, 0 < h < \delta_0$, so that
$$|T_x \cap (x + (1 - 2\rho)h, x + (1 - \rho)h)|^e > \frac{\rho h}{m}.$$
This is possible because the right exterior density of T_x at x exceeds $1/m$ and so lemma $(A_{6.7})$ (from the appendix) may be applied. We choose then a point y in the set
$$Y \cap (x + (1 - \rho)h, x + h),$$
which point must exist because of the porosity assumption on the set Y.

We obtain our contradiction by showing that such a point y cannot exist. Because of the density requirements on T_x and S_y, and the nature of the partition we have con-

structed there must be a point z in the intersection
$$T_x \cap S_y \cap (x + (1 - 2\rho)h, x + (1-\rho)h).$$
For such a point we must have
$$\frac{f(z) - f(x)}{z - x} > r, \quad \frac{f(y) - f(z)}{y - z} > p,$$
$$z - x > (1 - 2\rho)h \quad \text{and} \quad y - z < 2\rho h.$$
We put these together to obtain
$$f(y) - f(x) = f(z) - f(x) + f(y) - f(z)$$
$$\geq r(z - x) + p(y - z) = r(y - x) - (r - p)(y - z)$$
$$\geq \left[r - \frac{(r - p)2\rho}{1 - 2\rho}\right](y - x).$$
On the other hand the density requirements on S_y and S_x allow us to conclude that there is a point z' in the intersection
$$S_y \cap S_x \cap (x, x - \frac{2\rho(y - x)}{1 - 2\rho}),$$
and from this we obtain the inequalities
$$\frac{f(x) - f(z')}{x - z'} > p \quad \text{and} \quad \frac{f(y) - f(z')}{y - z'} < q.$$
From this we may conclude that
$$f(y) - f(x) = f(y) - f(z') - [f(x) - f(z')]$$
$$\leq q(y - z') - p(x - z') = q(y - x) + (q - p)(x - z')$$
$$\leq \left[q + \frac{(q - p)2\rho}{1 - 2\rho}\right](y - x).$$
Putting these inequalities together we must have
$$r - \frac{2\rho(r - p)}{1 - 2\rho} \leq \frac{f(y) - f(x)}{y - x} \leq q + \frac{(q - p)2\rho}{1 - 2\rho},$$
and this contradicts the choice of the number ρ made previously. From this contradiction we obtain the porosity result required and the theorem is proved.

(72.3) THEOREM. For an arbitrary function f the set of points x where
$$\overline{D}_{ap}{}^+ f(x) < \overline{D}_{ap}{}^- f(x) < +\infty \quad \text{or} \quad \overline{D}_{ap}{}^- f(x) < \overline{D}_{ap}{}^+ f(x) < +\infty$$
is σ-porous.

PROOF. The proof is obtained similarly to that of the preceding theorem but with the appropriate computational changes. Again let us define the set of points
$$X_{pqr} = \{x : \overline{D}_{ap}{}^- f(x) < p < q < \overline{D}_{ap}{}^+ f(x) < r\}$$
for fixed rational numbers p, q, and r. By our usual devices the theorem is proved if we are able to show that this set is σ-porous.

For each $x \in X_{pqr}$ define the sets
$$S_x = \{y : \frac{f(y) - f(x)}{y - x} < p\}$$
$$T_x = \{y : \frac{f(y) - f(x)}{y - x} < r\}$$
and
$$U_x = \{y : \frac{f(y) - f(x)}{y - x} > q\}.$$

Note that the set S_x must have (interior) density 1 on the left at x, that T_x must have (interior) density 1 on the right at x, and that the set U_x must have (exterior) upper density positive on the right at x. For each integer m we may collect the set of points X_{pqrm} at which this latter upper density exceeds $1/m$. Using the density requirements above we may define a guage δ on X_{pqrm} so that if $0 < t < 2\delta(x)$ then

$$\left|(x, x-t) \setminus S_x\right|^e < \frac{\rho t}{2m}$$

and

$$\left|(x, x+t) \setminus T_x\right|^e < \frac{\rho t}{2m},$$

where ρ is a number from the interval $(0, 1/2)$ taken so that

$$\frac{2\rho}{1-2\rho} < \frac{q-p}{(r-q)+(r-p)}.$$

This guage δ induces a denumerable partition of the set X_{mrsp}, which partition we may simply denote by $\{Y_n\}$. We claim that the right porosity of each set Y_n at each of its points must exceed $\rho/(1-\rho)$. In order to obtain a contradiction let us suppose that there is a point x_0 in the set Y_n such that the right porosity of Y_n at x_0 is less than $\rho/(1-\rho)$. Then there is a positive number δ_0 such that

$$\lambda(Y_n, x_0, x_0+t) < \frac{\rho t}{1-\rho} \quad \text{if } 0 < t < \delta_0.$$

Choose a number h, $0 < h < \delta_0$, so that

$$\left|U_x \cap (x+(1-\rho)h, x+h)\right|^e > \frac{\rho h}{m}.$$

This is possible because the right exterior density of U_x at x exceeds $1/m$ and so, as before, lemma $(A_{6.7})$ may be applied. We choose then a point y in the set

$$Y \cap (x+(1-2\rho)h, x+(1-\rho)h),$$

which point must exist because of the porosity assumption on the set Y.

We obtain our contradiction by showing that such a point y cannot exist. Indeed consider the set U_y. Because of the density requirements on U_x and T_y, and the nature of the partition we have constructed there must be a point z in the intersection

$$U_x \cap T_y \cap (x+(1-\rho)h, x+h).$$

For such a point we must have

$$\frac{f(z)-f(x)}{z-x} > q, \quad \frac{f(z)-f(y)}{z-y} < r,$$
$$z - y < 2\rho h \quad \text{and} \quad y - x > (1-2\rho)h.$$

Together these give

$$f(y) - f(x) = f(z) - f(x) - [f(z) - f(y)]$$
$$\geq q(z-x) - r(z-y) = q(y-x) - (r-q)(z-y)$$
$$= \left[q - \frac{(r-q)2\rho}{1-2\rho}\right](y-x).$$

On the other hand the density requirements on S_y and T_x allow us to conclude that there is a point z' in the intersection

$$S_y \cap T_x \cap (x, x + \frac{2\rho(y-x)}{1-2\rho}),$$

and from this we obtain the inequalities

$$\frac{f(z')-f(x)}{z'-x} < r \quad \text{and} \quad \frac{f(y)-f(z')}{y-z'} < p.$$

From this we may conclude that

$$f(y) - f(x) = f(y) - f(z') + f(z') - f(x)$$
$$\leq p(y - z') + r(z' - x) = p(y - x) + (r - p)(z' - x)$$
$$= \left[p + \frac{(r-p)2\rho}{1 - 2\rho}\right](y - x).$$

Putting these inequalities together we must have

$$q - \frac{2\rho(r-q)}{1 - 2\rho} \leq \frac{f(y) - f(x)}{y - x} \leq p + \frac{(r-p)2\rho}{1 - 2\rho},$$

and this contradicts the choice of the number ρ made previously. From this contradiction we obtain the porosity result required and the theorem is proved.

Let us observe that, because of the last two theorems, we may state another version of the Denjoy-Young relations for the approximate Dini derivatives. For an arbitrary function f one must have at every point x, with the possible exception of x in a σ-porous set, one of the following three situations: either

$$\overline{D}_{ap}{}^+ f(x) = \overline{D}_{ap}{}^- f(x) \quad \text{and} \quad \underline{D}_{ap}{}^- f(x) = \underline{D}_{ap}{}^+ f(x),$$

or

$$\overline{D}_{ap}{}^+ f(x) = +\infty \quad \text{and} \quad \underline{D}_{ap}{}^- f(x) = -\infty,$$

or

$$\underline{D}_{ap}{}^+ f(x) = -\infty \quad \text{and} \quad \overline{D}_{ap}{}^- f(x) = +\infty.$$

This result is due to Preiss and Zajíček.

These results are balanced by the following examples.

(72.4) **Example.** There is a continuous function f so that the set of points x at which

$$\overline{D}_{ap}{}^+ f(x) = \overline{D}_{ap}{}^- f(x) = +\infty$$

and

$$\underline{D}_{ap}{}^+ f(x) = \underline{D}_{ap}{}^- f(x) = -\infty$$

is residual. Indeed take any continuous function f for which $\overline{D}^{\#} f(x) = +\infty$ and $\underline{D}^{\#} f(x) = -\infty$ everywhere and apply Theorem (64.1).

(72.5) **Example.** Given any numbers $\alpha \leq \beta$ there is a continuous function f for which the set of points x at which

$$\overline{D}_{ap}{}^+ f(x) = \overline{D}_{ap}{}^- f(x) = \beta$$

and

$$\underline{D}_{ap}{}^+ f(x) = \underline{D}_{ap}{}^- f(x) = \alpha$$

is residual. This is given in Zajíček [272, p.559].

(72.6) **Example.** Preiss and Zajíček [198, pp.694-697] give a more difficult construction. They show that for any numbers α and β there is a function f so that the set of points x at which

$$\overline{D}_{ap}{}^+ f(x) = \alpha, \quad \overline{D}_{ap}{}^- f(x) = +\infty,$$
$$\underline{D}_{ap}{}^+ f(x) = -\infty \quad \text{and} \quad \underline{D}_{ap}{}^- f(x) = \beta$$

is residual.

§73. σ-porous version.

One interesting variant on the DYS theorem was obtained, independently, by Zajíček [270] and by Belna, Cargo, Evans and Humke [11]. The problem posed in these two articles was to obtain a set of relations for the ordinary Dini derivatives similar to those for the original DYS theorem but for which the exceptional set would be sharper than measure zero, would be in fact σ-porous. Their version reads as follows.

For an arbitrary function f, at every point x, with the exception only of a set that is σ-porous, one of the following three situations must occur: either

$$\underline{D}^- f(x) = \underline{D}^+ f(x) \text{ and } \overline{D}^- f(x) = \overline{D}^+ f(x)$$

or

$$-\infty = \underline{D}^- f(x) \leq \underline{D}^+ f(x) \leq \overline{D}^- f(x) \leq \overline{D}^+ f(x) = +\infty,$$

or

$$-\infty = \underline{D}^+ f(x) \leq \underline{D}^- f(x) \leq \overline{D}^+ f(x) \leq \overline{D}^- f(x) = +\infty.$$

This result follows from the following two elementary theorems which are similar in statement and proof to the material in the preceding sections. Note that both of the papers cited above (Zajíček and Belna et al.) use the Blumberg-Jarník method rather than the less elegant, but more direct method that we employ here.

(73.1) THEOREM. For an arbitrary function f the sets of points

$$\{ x : \overline{D}^- f(x) < \overline{D}^+ f(x) < +\infty \},$$

and

$$\{ x : \overline{D}^+ f(x) < \overline{D}^- f(x) < +\infty \}$$

are σ-porous.

PROOF. We obtain that the first of these sets is σ-porous by showing that for any rational numbers $r, s, (r < s)$ and positive integer m the set

$$X = \{ x : \overline{D}^- f(x) < r < s < \overline{D}^+ f(x) < m \}$$

is σ-porous. Since the exceptional set of the theorem is a denumerable union of such sets this provides a proof.

For each $x \in X$ define the sets

$$S_x = \{ y : y = x \text{ or } \frac{f(y) - f(x)}{y - x} < r \}$$

and

$$T_x = \{ y : y = x \text{ or } \frac{f(y) - f(x)}{y - x} < m \}.$$

By the nature of the set X this requires that there is a guage δ on X so that

$$S_x \supset (x - \delta(x), x] \text{ and } T_x \supset [x, x + \delta(x)).$$

Let $\{ X_n \}$ be the partition of X induced by the guage δ. We claim that the right porosity of X_n at any one of its points must exceed $p = (s - r)/m$. From this it follows that each X_n is porous and so X is σ-porous, as required.

Suppose contrary to this that there is a point x_0 in some set X_n at which the right porosity is less than $p = (s - r)/m$, so that a positive number δ_0 may be found such that

$$\lambda(X_n, x_0, x_0 + t) < pt \quad \text{for } 0 < t < \delta_0.$$

Let x_1 be any point in the interval $(x_0, x_0 + \delta_0)$; we obtain the inequality

$$\frac{f(x_1) - f(x_0)}{x_1 - x_0} \leq s. \tag{*}$$

To see this let $h = x_1 - x_0$ and use the porosity estimate above to obtain a point x_2 from the set

$$X_n \cap (x_1 - ph, x_1).$$

This gives

$$f(x_2) - f(x_0) < r(x_2 - x_0) \quad \text{and} \quad f(x_1) - f(x_2) < m(x_1 - x_2)$$

and hence that

$$f(x_1) - f(x_0) < r(x_2 - x_0) + m(x_1 - x_2) \leq rh + mph \leq h(r + \frac{m(s-r)}{m}) \leq s(x_1 - x_0)$$

which is precisely (*). But this inequality contradicts the fact that

$$\overline{D}^+ f(x_0) > s$$

and this contradiction completes the proof.

(73.2) THEOREM. For an arbitrary function f the sets of points

$$\{ x : -\infty < \underline{D}^- f(x) \leq \overline{D}^- f(x) < +\infty \quad \text{and} \quad \overline{D}^+ f(x) = +\infty \}$$

and

$$\{ x : -\infty < \underline{D}^+ f(x) \leq \overline{D}^+ f(x) < +\infty \quad \text{and} \quad \overline{D}^- f(x) = +\infty \}$$

are σ-strongly-porous.

PROOF. Let us address just the latter of these sets. As usual it is sufficient that we consider each set

$$X_m = \{ x : -m < \underline{D}^+ f(x) \leq \overline{D}^+ f(x) < m, \overline{D}^- f(x) = +\infty \}$$

for integers m, and prove that each such set is σ-strongly-porous. We may define a guage δ on X_m so that

$$\left| \frac{f(x) - f(y)}{y - x} \right| < m \quad \text{if } 0 < y - x < \delta(x), \ x \in X_m.$$

The guage δ induces a denumerable partition $\{X_{mn}\}$ of the set X_m and we shall obtain our result by proving that each member of this partition is strongly porous. In order to obtain a contradiction let us suppose, contrary to this, that there is a set X_{mn} and a point $x_0 \in X_{mn}$ at which the left porosity index is not $+\infty$, but is less than some number $t < +\infty$. Thus there must be a sequence of points $\{x_k\}$ in X_{mn} with $x_k < x_0$ so that $x_k \to x_0$ and

$$x_{k+1} - x_k < t(x_0 - x_{k+1}).$$

Since each point x_k belongs to X_{mn} we have by the nature of the partition that for any $x_k \leq y \leq x_{k+1}$

$$|f(y) - f(x_k)| < m(y - x_k),$$
$$|f(x_{k+1}) - f(x_k)| < m(x_{k+1} - x_k),$$

and

$$|f(x_0) - f(x_{k+1})| < m(x_0 - x_{k+1}).$$

Putting these together we obtain

$$|f(x_0) - f(y)| \le |f(y) - f(x_k)| + |f(x_{k+1}) - f(x_k)| + |f(x_0) - f(x_{k+1})|$$
$$< m\{(y - x_k) + (x_{k+1} - x_k) + (x_0 - x_{k+1})\}$$
$$< m\{(x_0 - y) + 2(x_{k+1} - x_k)\}$$
$$< m[x_0 - y]\left[1 + 2\,\frac{x_{k+1} - x_k}{x_0 - x_{k+1}}\right].$$
$$< m[x_0 - y](1 + 2t).$$

But this holds, then, for all $x_1 < y < x_0$ which places an upper bound of $m(1+2t)$ on $\overline{D}^- f(x_0)$, and contradicts the fact that this is to be $+\infty$. This contradiction proves the theorem.

§74. The sharp version.

The theorem of W.H. Young that we proved in section 64 may be also expressed as a variant of the DYS theorem. If we interpret the result in corollary (64.3) in the appropriate fashion we obtain the following result (which is pointed out in Belna, Cargo, Evans and Humke [11]).

For an arbitrary function f, at every point x, with the exception only of a set of the first category, one of the following three situations must occur: either

$$\underline{D}^- f(x) = \underline{D}^+ f(x) = \underline{D}^\# f(x) \text{ and } \overline{D}^- f(x) = \overline{D}^+ f(x) = \overline{D}^\# f(x),$$

or

$$-\infty = \underline{D}^- f(x) \le \underline{D}^+ f(x) \le \overline{D}^- f(x) \le \overline{D}^+ f(x) = +\infty,$$

or

$$-\infty = \underline{D}^+ f(x) \le \underline{D}^- f(x) \le \overline{D}^+ f(x) \le \overline{D}^- f(x) = +\infty.$$

This may be proved directly from the result mentioned above and the G.C. Young theorem.

The articles of Belna et al. [11], Zajíček [272], and Preiss and Zajíček [198] give some examples to show that the representation given above may not be improved in certain ways. Let us list these here.

(74.1) Example. There exists a continuous strictly increasing function f so that the derivative $f'(x) = +\infty$ for a residual set.

Thus the first of the above sets (say C_1) may not be replaced by the union of the sets of points C_{11} where

$$f'(x) \text{ exists (finitely)}$$

and the set C_{12} where

$$\underline{D}^- f(x) = \underline{D}^+ f(x) = -\infty \quad \text{and} \quad \overline{D}^- f(x) = \overline{D}^+ f(x) = +\infty.$$

A construction may be found, for example, in Natanson [169, pp. 214–215].

(74.2) Example. There exists a continuous strictly increasing function f so that
$$\underline{D}^- f(x) = \underline{D}^+ f(x) = \underline{D}^\# f(x) = 0$$
and

$$\overline{D}^- f(x) = \overline{D}^+ f(x) = \overline{D}^\# f(x) = 1$$

for a residual set.

Thus again the set of points C_1 above may not be replaced by the union of the three sets C_{11}, C_{12}, and C_{13} where C_{11} and C_{12} are as in the preceding example and C_{13} is the set of points where

$$\underline{D}^- f(x) = \underline{D}^+ f(x) \text{ (finite)} \quad \text{and} \quad \overline{D}^- f(x) = \overline{D}^+ f(x) = +\infty .$$

For the construction see Belna et al. [11, Ex.2].

(74.3) Example. Either of the sets of points
$$-\infty = \underline{D}^- f(x) < \underline{D}^+ f(x) < \overline{D}^- f(x) < \overline{D}^+ f(x) = +\infty ,$$
or
$$-\infty = \underline{D}^+ f(x) < \underline{D}^- f(x) < \overline{D}^+ f(x) < \overline{D}^- f(x) = +\infty$$
may be residual. Again see the article of Belna et al.; their construction gives f bounded and in the first class of Baire. Recall that, by the W.H.Young theorem, such a function could not be continuous.

§75. DYS for the negligent Dini derivatives.

Completely analogous results as we have seen in section 78 for the approximate Dini derivatives may be obtained for the negligent Dini derivatives. The first such study seems to have been done by Császár [52]. In the special case of the qualitative Dini derivatives the articles of Evans [68], [69] and Evans and Larson [75] contain a number of observations.

(75.1) THEOREM. For an arbitrary function f the set of points x where
$$-\infty < \underline{D_N}^+ f(x) \leq \overline{D_N}^+ f(x) < \overline{D_N}^- f(x) ,$$
or
$$-\infty < \underline{D_N}^- f(x) \leq \overline{D_N}^- f(x) < \overline{D_N}^+ f(x)$$
is σ-porous.

PROOF. The proof is almost identical with that for Theorem (72.2). As usual we define the set of points
$$X_{pqr} = \{ x: p < \underline{D_N}^- f(x) \leq \overline{D_N}^- f(x) < q < r < \overline{D_N}^+ f(x) \}$$
for fixed rational numbers p, q, and r. Clearly the theorem is proved if we are able to show that this set is σ-porous.

For each $x \in X_{pqr}$ define the sets
$$S_x = \{ y: p < \frac{f(y) - f(x)}{y - x} < q \}$$
and
$$T_x = \{ y: \frac{f(y) - f(x)}{y - x} > r \} .$$
We may define a guage δ on X_{pqr} so that
$$(x - t, x) \setminus S_x \in N \quad \text{if} \quad 0 < t < 2\delta(x) .$$

This guage δ induces a denumerable partition of the set X_{pqr}, which partition we may denote by $\{Y_n\}$. Let ρ denote a number from the interval $(0, 1/2)$ chosen so that

$$\frac{2\rho}{1 - 2\rho} < \frac{r - q}{(r - p) + (q - p)}.$$

We claim that the right porosity of each set $Y = Y_n$ at each of its points must be at least ρ. In order to obtain a contradiction let us suppose that there is a point x_0 in the set Y such that the right porosity of Y at x_0 is less than ρ, and hence there is a number δ_0 with

$$\lambda(Y, x_0, x_0 + t) < \rho t \quad \text{if } 0 < t < \delta_0.$$

Choose a number h, $0 < h < \delta_0$, so that

$$T_x \cap (x + (1 - 2\rho)h, x + (1 - \rho)h)$$

does not belong to the σ-ideal N. This is possible because of the following argument. We know by the construction that $T_x \cap (x, x + t)$ cannot belong to N for any $t > 0$. Thus for any θ, $0 < \theta < 1$, and any $h > 0$ there must be some n for which

$$T_x \cap (x + \theta^{n+1}h, x + \theta h)$$

does not belong to N otherwise

$$T_x \cap (x, x + \theta h) = \sum_{n=0}^{\infty} T_x \cap (x + \theta^{n+1}h, x + \theta h)$$

would belong to N which is impossible.

We choose then a point y in the set

$$Y \cap (x + (1 - \rho)h, x + h),$$

which point must exist because of the porosity assumption on the set Y. We obtain our contradiction by showing that such a point y cannot exist. Because of the requirements on T_x and S_y, and the nature of the partition we have constructed there must be a point z in the intersection

$$T_x \cap S_y \cap (x + (1-2\rho)h, x + (1 - \rho)h).$$

For such a point we must have

$$\frac{f(z) - f(x)}{z - x} > r, \quad \frac{f(y) - f(z)}{y - z} > p,$$

$$z - x > < (1 - 2\rho)h \quad \text{and} \quad y - z < 2\rho h.$$

Putting these together we have

$$f(y) - f(x) = f(z) - f(x) + f(y) - f(z)$$
$$\geq r(z - x) + p(y - z) = r(y - x) - (r - p)(y - z)$$
$$\geq \left[r - \frac{(r - p)2\rho}{1 - 2\rho}\right](y - x).$$

On the other hand the requirements on S_y and S_x allow us to conclude that there is a point z' in the intersection

$$S_y \cap S_x \cap (x, x - \frac{2\rho(y - x)}{1 - 2\rho}),$$

and from this we obtain the inequalities

$$\frac{f(x) - f(z')}{x - z'} > p \quad \text{and} \quad \frac{f(y) - f(z')}{y - z'} < q.$$

From this we may conclude that

$$f(y) - f(x) = f(y) - f(z') - [f(x) - f(z')]$$

$$< q(y-z') - p(x-z') = q(y-x) + (q-p)(x-z')$$
$$\leq \left[q + \frac{(q-p)2\rho}{1-2\rho}\right](y-x).$$

Putting the inequalities together we must have
$$r - \frac{2\rho(r-p)}{1-2\rho} \leq \frac{f(y)-f(x)}{y-x} \leq q + \frac{(q-p)2\rho}{1-2\rho},$$
and this contradicts the choice of the number ρ made previously. From this contradiction we obtain the porosity result required and the theorem is proved.

(75.2) THEOREM. For an arbitrary function f the set of points x where
$$\overline{D}_N^+ f(x) < \overline{D}_N^- f(x) < +\infty \quad \text{or} \quad \overline{D}_N^- f(x) < \overline{D}_N^+ f(x) < +\infty$$
is σ-porous.

PROOF. The proof is obtained similarly to that of the preceding theorem and almost identical with the proof of Theorem (72.3). Again let us define the set of points
$$X_{pqr} = \{x: \overline{D}_N^- f(x) < p < q < \overline{D}_N^+ f(x) < r\}$$
for fixed rational numbers $p, q,$ and r. By our usual devices the theorem is proved if we are able to show that this set is σ-porous.

For each $x \in X_{pqr}$ define the sets
$$S_x = \{y: \frac{f(y)-f(x)}{y-x} < p\}$$
$$T_x = \{y: \frac{f(y)-f(x)}{y-x} < r\}.$$
and
$$U_x = \{y: \frac{f(y)-f(x)}{y-x} > q\}.$$

We may define a guage δ on X_{pqr} so that if $0 < t < 2\delta(x)$ then
$$(x, x-t) \setminus S_x \in N \quad \text{and} \quad (x, x+t) \setminus T_x \in N.$$
Again let ρ be a number from the interval $(0, 1/2)$ taken so that
$$\frac{2\rho}{1-2\rho} < \frac{q-p}{(r-q)+(r-p)}.$$

This guage δ induces a denumerable partition of the set X_{pqr}, which partition we may simply denote by $\{Y_n\}$. We claim that the right porosity of each set Y_n at each of its points must exceed $\rho/(1-\rho)$. In order to obtain a contradiction let us suppose that there is a point x_0 in the set Y_n such that the right porosity of Y_n at x_0 is less than $\rho/(1-\rho)$. Then there is a positive number δ_0 such that
$$\lambda(Y_n, x_0, x_0+t) < \frac{\rho t}{1-\rho} \quad \text{if} \quad 0 < t < \delta_0.$$
Choose a number $h, 0 < h < \delta_0$, so that the set
$$U_x \cap (x+(1-\rho)h, x+h)$$
does not belong to the σ-ideal. The same argument given in the proof of the preceding theorem supplies this fact. We choose then a point y in the set
$$Y \cap (x+(1-2\rho)h, x+(1-\rho)h),$$
which point must exist because of the porosity assumption on the set Y.

We obtain our contradiction by showing that such a point y cannot exist. Indeed consider the set U_y. Because of the requirements on U_x and T_y, and the nature of the

partition we have constructed there must be a point z in the intersection
$$U_x \cap T_y \cap (x + (1-\rho)h, x+h).$$
For such a point we must have
$$\frac{f(z) - f(x)}{z - x} > q, \quad \frac{f(z) - f(y)}{z - y} < r,$$
$$z - y < 2\rho h \quad \text{and} \quad y - x > (1 - 2\rho)h.$$
Putting these together we find
$$f(y) - f(x) = f(z) - f(x) - [f(z) - f(y)]$$
$$\geq q(z-x) - r(z-y) = q(y-x) - (r-q)(z-y)$$
$$= \left[q - \frac{(r-q)2\rho}{1 - 2\rho} \right](y - x).$$
On the other hand the requirements on S_y and T_x allow us to conclude that there is a point z' in the intersection
$$S_y \cap T_x \cap (x, x + \frac{2\rho(y-x)}{1 - 2\rho}),$$
and from this we obtain the inequalities
$$\frac{f(z') - f(x)}{z' - x} < r \quad \text{and} \quad \frac{f(y) - f(z')}{y - z'} < p.$$
From this we may conclude that
$$f(y) - f(x) = f(y) - f(z') + f(z') - f(x)$$
$$\leq p(y - z') + r(z' - x) = p(y-x) + (r-p)(z'-x)$$
$$= \left[p + \frac{(r-p)2\rho}{1 - 2\rho} \right](y - x).$$
Putting the inequalities together we must have
$$q - \frac{2\rho(r-q)}{1 - 2\rho} \leq \frac{f(y) - f(x)}{y - x} \leq p + \frac{(r-p)2\rho}{1 - 2\rho},$$
and this contradicts the choice of the number ρ made previously. From this contradiction we obtain the porosity result required and the theorem is proved.

These results permit us to state another version of the Denjoy-Young relations, this time for the negligent Dini derivatives. For an arbitrary function f one must have at every point x, with the possible exception of x in a σ-porous set, one of the following three situations: either
$$\overline{D}_N^+ f(x) = \overline{D}_N^- f(x) \quad \text{and} \quad \underline{D}_N^- f(x) = \underline{D}_N^+ f(x),$$
or
$$\overline{D}_N^+ f(x) = +\infty \quad \text{and} \quad \underline{D}_N^- f(x) = -\infty,$$
or
$$\underline{D}_N^+ f(x) = -\infty \quad \text{and} \quad \overline{D}_N^- f(x) = +\infty.$$
In the special case where N is taken as the σ-ideal of first category sets this was first proved by Evans [68].

APPENDIX

SET POROSITY

§A₁. Introduction.

In this appendix we collect a variety of computations directly related to the notion of set porosity. As yet this material may only be found scattered in the literature and it takes a determined effort for the novice to find and study these ideas. It is hoped that this material will be of some use to the reader.

In addition let us mention the following further sources of information on the subject of set porosity. The structure of porous sets and σ-porous sets has been studied by Foran and Humke [78], Humke and Vessey [123], and by Tkadlec [240]. The notion of set porosity plays an implicit role in a number of instances in analysis; most notably the classification of sets introduced by Zahorski [265] requires a number of porosity computations. Explicitly porosity has appeared as a tool in several other more recent papers. The articles of Belna, Cargo, Evans, and Humke [11], Belna, Evans and Humke [12], Bruckner and Haussermann [38], Bruckner and Thomson [44], Bruckner, Laczkovich, Petruska and Thomson [41], Haussermann [109], [110], Humke and Preiss [121], Humke and Thomson [122], Thomson [236], [239], Tkadlec [240], [241], Vessey [244], and Zajíček [268], [269], [270], [271] can be consulted for a number of diverse applications of the concept.

§A₂. Basic definitions.

The notion of set porosity first appeared (under a different nomenclature) in some early works of Denjoy [62], [64] and Khintchine [136] and then arose independently in the study of cluster sets in 1967 (Dolženko [67]). Although the basic computations are the same in these sources the intention is quite different. Denjoy was interested in obtaining a classification of perfect sets on the real line in terms of the relative sizes of the complementary intervals; Khintchine had required a convenient way of describing certain arguments that use density considerations. On the other hand the cluster set theorists required a general way of describing certain exceptional sets (the σ-porous sets) that had arisen in a number of investigations and which form a subclass of the class of measure-zero, first category sets. Note the distinctions: the former authors needed certain geometric properties of sets, while the cluster set theorists required only the description of a σ-ideal that arises naturally in certain problems. Since the reintroduction of these ideas by the cluster theorists a number of real analysts have shown the role that the idea can play in numerous questions, both in the local sense and in the global sense.

We use some of the terminology and notation introduced by Dolženko but we prefer in a number of instances the formulations of Denjoy.

($A_{2}.1$) DEFINITION. Let E be a set and let $a < b$. Then we write $\lambda(E, a, b)$ and $\lambda(E, b, a)$ for the length of the largest open subinterval of (a, b) that contains no point of E.

Since $\lambda(E, a, b) = \lambda(\overline{E}, a, b)$ this notion could well have been restricted to closed sets E but that would be a notational inconvenience; it is also convenient to be able to write $\lambda(E, x, y)$ both for $x < y$ and $y < x$.

($A_{2}.2$) DEFINITION. Let E be a set; then at any point x one defines the following porosity computations: the <u>right hand porosity</u> of E at x is defined as

$$p^{+}(E; x) = \limsup_{h \to 0+} \frac{\lambda(E, x, x+h)}{h};$$

the <u>left hand porosity</u> of E at x is defined as

$$p^{-}(E; x) = \limsup_{h \to 0+} \frac{\lambda(E, x, x-h)}{h};$$

and the <u>bilateral porosity</u> of E at x is defined as

$$p(E; x) = \limsup_{h \to 0} \frac{\lambda(E, x, x+h)}{|h|},$$

which latter is, of course, the maximum of the two unilateral porosities.

Note that the porosity (bilateral, right, or left) is always a number in the interval $[0, 1]$ and that both extremes can occur. It is convenient to have a suggestive language that allows one to discuss the porosity of a set in an informal style. Thus the following terminology has evolved. Locally one says that a set E is

($A_{2}.3$) <u>porous</u> at x if $p(E; x) > 0$,

($A_{2}.4$) <u>porous on the right</u> at x if $p^{+}(E; x) > 0$.

($A_{2}.5$) <u>nonporous</u> at x if $p(E; x) = 0$,

($A_{2}.6$) <u>strongly porous</u> at x if
$$p^{+}(E; x) = 1 \quad \text{or} \quad p^{-}(E; x) = 1,$$
and

($A_{2}.7$) <u>bilaterally strongly porous</u> at x if
$$p^{+}(E; x) = p^{-}(E; x) = 1.$$

Globally one says that a set E is

($A_{2}.8$) <u>porous</u> if it is porous at each of its points,

($A_{2}.9$) <u>strongly porous</u> if it is strongly porous at each of its points,

($A_{2}.10$) <u>σ-porous</u> if it can be expressed as a countable union of porous sets, and finally

(A$_{2}$.11) σ-strongly porous if it can be expressed as a countable union of strongly porous sets.

These notions all derive from the simple idea of measuring the relative sizes of the "holes" in the set E; Denjoy considered this same computation and showed that it was equivalent to another way of looking at these holes. One considers instead sequences $\{x_n\}$ contained in E and converging to a point x. Then the limits of the ratios

$$\frac{x_n - x}{x_{n+1} - x}$$

again give an indication of the size of the holes in E near to x. Estimates of this type arise frequently in the study of derivatives and it is useful to have a language for them; for this reason we introduce a version of Denjoy's index and show how it is related to the porosity computation.

(A$_{2}$.12) DEFINITION. Let E be a set and x a point. Then the <u>right porosity index</u> of E at x, $PI^+(E;x)$ is defined to be the supremum of all numbers r for which a sequence of intervals $\{I_n\}$ converging to x on the right may be found such that each I_n is disjoint from E and

$$r < \frac{|I_n|}{d(x, I_n)}$$

(where throughout $d(x, I)$ denotes the distance from the point x to the interval I).

Following the usual convention we write $PI^+(E;x) = 0$ in the event that no such number r exists. The left porosity index $PI^-(E;x)$ is similarly defined. This computation may also be expressed by writing $PI^+(E;x)$ as

$$\lim_{\varepsilon \to 0+} \sup \left\{ \frac{k}{h} : (x+h, x+h+k) \cap E = \emptyset, h + k < \varepsilon, 0 \leq h, 0 < k \right\}.$$

This index computation differs slightly from that used by Denjoy but it is merely a rescaling of his index; indeed as we shall see in Lemma (A$_{2}$.13) this index is just a rescaling of the porosity itself. Although Denjoy uses the "index" computation rather than the closely related porosity computation, the notion of porosity, at least for one dimensional sets, should be attributed to him. (He did not, however, exploit the notion of a σ-porous set.) The exact relation between porosity and the porosity index is given in the next lemma which is essentially due to Denjoy.

(A$_{2}$.13) LEMMA. Let E be a set and x a point. If $p = p^+(E;x)$ is the right hand porosity of E at x and $i = PI^+(E;x)$ is the right porosity index of E at x then one must have the identities

$$i = \frac{p}{1-p} \quad \text{and} \quad p = \frac{i}{i+1}.$$

The same identity holds for the left porosity and the left porosity index.

PROOF. The proof follows from some elementary arithmetic. The porosity p may be computed as the limit superior of the expression

$$|I_n| / (d(x, I_n) + |I_n|)$$

for sequences of intervals I_n converging to x on the right, with each interval I_n disjoint from the set E. Similarly the index i is computed as a limit superior, for such sequences, of the expression $|I_n|/ d(x, I_n)$. But

$$\frac{|I|}{d(x,I)} = \left(\frac{|I|}{d(x,I) + |I|}\right)\left(1 - \frac{|I|}{d(x,I) + |I|}\right)^{-1},$$

and this gives $i = p(1-p)^{-1}$.

As a result of this lemma we see that the porosity and the porosity index are just different scalings of the same estimate. The porosity measure is a number in the interval $[0, 1]$ with porosity 0 (nonporous) for the fattest sets and porosity 1 (strongly porous) for the thinnest sets; the porosity index merely rescales this to the interval $[0, +\infty]$ with index 0 corresponding to nonporous and index $+\infty$ corresponding to strongly porous. It is convenient to have both expressions available; in particular for generalizations (see section A_7 below) we prefer to formulate the generalization from the porosity index.

The next two lemmas show how the porosity may be viewed from slightly different perspectives. The first addresses the porosity at a point x from the notion of how slowly one may have a sequence of points in the set converge to x. The second shows how porosity computations can be made on the basis of the relative size of the complementary intervals. Both of these ideas may be found in the original papers of Denjoy. The proofs are elementary.

(A_2.14)   LEMMA. The right porosity index of a set E at a point x is the infimum of all numbers c such that a decreasing sequence of points $\{x_n\}$ belonging to E may be found so that $x_n \to x$ and

$$(1 + c)(x_{n+1} - x) < (x_n - x).$$

(A_2.15)   LEMMA. Let P be a perfect nowhere dense set and let $\{(a_n, b_n)\}$ be the sequence of intervals complementary to P. Then at any point $x \in P$ and not isolated on the right in P,

$$p^+(P; x) = \limsup \frac{b_n - a_n}{b_n - x}$$

and

$$PI^+(P; x) = \limsup \frac{b_n - a_n}{a_n - x}$$

as a_n, b_n converge to x, with $a_n > x$.

There is another way to use this observation which is notationally convenient. The porosity computations here state the relative sizes that the complementary intervals to the set P may have. One can view this as asserting that, should these intervals be enlarged by some factor, they would necessarily include points at which the porosity may be specified. To make this notion precise let $I = (a, b)$ be an interval and let ψ be a

positive function defined on $(0, +\infty)$. We define the interval $EXP(I, \psi)$ to be that interval concentric with I and with length
$$|I| + 2\psi(|I|),$$
that is to say
$$EXP(I, \psi) = (c, d)$$
where
$$c = a - \psi(b-a) \quad \text{and} \quad d = b + \psi(b-a).$$
Thus $EXP(I, \psi)$ is just the interval I enlarged by attaching to each side an interval of length $\psi(|I|)$. Most frequently ψ is taken as a linear function $\psi(t) = \alpha t$ ($\alpha > 0$) in which case we would write merely $EXP(I, \alpha)$. Note that for α close to zero the interval $EXP(I, \alpha)$ is a slight enlargement of the interval I.

In the literature (see for example Zajíček [269]) the notation $c*I$ is used to indicate an interval concentric with I and with length $c|I|$. As we shall use the porosity index computation rather than the porosity computation itself we shall find the $EXP(I, \alpha)$ notation more convenient.

The above lemma of Denjoy may now be expressed in the following useful manner.

($A_2.16$) LEMMA. Let P be a perfect set and I_n the sequence of open intervals complementary to P. Then if x is a point of P that is not isolated on right in P and which has right porosity index exceeding a positive number i then x must belong to infinitely many intervals
$$EXP(I_n, \frac{1}{i}) \quad n = 1, 2, 3, \ldots.$$
PROOF. This follows immediately.

This may also be expressed in the following equivalent form.

($A_2.17$) COROLLARY. Let P be a perfect set and I_n the sequence of open intervals complementary to P. Then, for each N, the set
$$\prod_{n=N}^{\infty} \sum_{k=n}^{\infty} EXP(I_k, c)$$
includes every point of P that is not isolated on either side and which has porosity index greater than $1/c$.

§A_3. Elementary properties.

We list in this section some elementary, but useful, properties of porosity as a local notion. In each case the proof is quite easy and will be omitted or briefly sketched.

($A_{3.1}$) LEMMA. The porosity or porosity index of a set at a point is identical with the corresponding notion for the closure of that set.

($A_{3.2}$) LEMMA. If the right [left] porosity of a set E at a point x exceeds a number q, $0 < q < 1$, then the right [left] lower density of E at x is at most $1 - q$.

PROOF. If the right porosity of E at x exceeds a number q, $0 < q < 1$, then there must be a sequence $\{h_n\}$ decreasing to zero so that
$$\lambda(E, x, x + h_n) > q h_n$$
for all n. Thus for this sequence
$$|E \cap (x, x + h_n)|^e \leq h_n - q h_n$$
and so the right lower (exterior) density of E at x is at most $1 - q$.

From this lemma the following four corollaries follow immediately.

($A_{3.3}$) COROLLARY. If the right porosity index of a set E at a point x exceeds the number c then the right lower density of E at x is at most $\frac{1}{c+1}$.

($A_{3.4}$) COROLLARY. At a point of density of a set the porosity is zero (i.e. every point of density is a point of nonporosity).

($A_{3.5}$) COROLLARY. At a point of right [left, bilateral] strong porosity of a set the right [left, bilateral] lower density must be zero.

($A_{3.6}$) COROLLARY. A porous set is nowhere dense and has measure zero.

($A_{3.7}$) LEMMA. Let E be an arbitrary set and let p be a positive number less than 1. Then the set of points x at which the right [left, bilateral] porosity of E at x is greater than p is a a set of type G_δ.

PROOF. The set of the lemma may be expressed as the intersection of the sequence of sets $\{G_n\}$ where
$$G_n = \{ x : \text{there exists } h \in (0, 1/n) \text{ with } \lambda(E, x, x + h) > ph \}.$$
One checks easily that each set G_n is open.

($A_{3.8}$) LEMMA. If P is perfect and nowhere dense then the set of points in P at which P is bilaterally strongly porous is residual in P.

PROOF. [cf. Denjoy [64, p.195]] For any integer n the set of points at which the right porosity exceeds $1 - 1/n$ is a G_δ subset of P by the lemma, so that the set of points at which the right porosity is exactly 1 must also be a G_δ. Since P is nowhere dense this set is dense in P and hence residual there. Similar arguments apply on the left and the lemma will then follow.

§A$_4$. σ-porous sets.

In this section we present some of the basic information on σ-porous sets. The notion itself is due to Dolženko and the basic results in the subject were worked out by Zajíček [269]. By lemma (A$_{3.6}$) we see that a σ-porous set is necessarily first category and measure zero. We state this as a lemma.

(A$_{4.1}$) LEMMA. The class of σ-porous sets is a σ-ideal contained in the class of measure zero sets and in the class of first category sets.

PROOF. That the collection of σ-porous sets is a σ-ideal of sets is easy to verify. That it is contained in the σ-ideal of first category, measure zero sets follows immediately from lemma (A$_{3.6}$). That there are sets in the latter σ-ideal that do not belong to the former will be proved in section A$_{11}$ below.

Our next lemma is a structural lemma for σ-porous sets due to Zajíček. This asserts that a σ-porous set A may be decomposed into a sequence of sets $\{A_n\}$ such that the porosity of each A_n at each of its points is relatively large.

(A$_{4.2}$) LEMMA. Let $c > 0$. Then any σ-porous set A may be expressed as the union of a sequence of sets $\{A_n\}$ such that the porosity index of each A_n at each of its points is at least c.

PROOF. The general method of proof is due to Zajíček [269]. We use this again in the proof of lemma (A$_{8.2}$), from which the present lemma could have been deduced. However the proof here is computationally simpler and the strategy easier to see.

It is clearly enough to suppose that A itself is porous; indeed we may suppose that the porosity index of A at each of its points exceeds a positive number ε. We also remove from A any point that is isolated on one side at least. Then if $\{I_i\}$ denotes the sequence of intervals complementary to A we know that for each $x \in A$ there are infinitely many indices i for which

$$d(x, I_i) < \varepsilon^{-1} |I_i|,$$

or, equivalently, for which

$$x \in EXP(I_i, \varepsilon^{-1}).$$

This fact is all that we need to know about such a set A in order to form the decomposition. Choose an integer M so large that $M\varepsilon > c$, and define the sets

$$C_m = A \cap \bigcap_{N=1}^{\infty} \sum_{n=N}^{\infty} EXP(I_i, mc^{-1}) \qquad m = 0, 1, 2, \ldots, M.$$

Here we consider that $EXP(I, 0) = I$. Note that $C_M = A$ because of the above porosity estimate on the points of A and lemma A$_{2.16}$. We show that each of the sets

$$C_1, \ C_2 \setminus C_1, \ C_3 \setminus C_2, \ \ldots, \ C_M \setminus C_{M-1}$$

may be expressed as the union of a sequence of sets each with the required porosity assertion (i.e. index at least 1). Let us write $C_0 = \emptyset$ and $EXP(I, 0) = I$. Then, for $m = 1, 2, \ldots, M$, we define

$$C_m \setminus C_{m-1} = \sum_{j=1}^{\infty} T_{mj}$$

where

$$T_{mj} = C_m \setminus \sum_{i=j}^{\infty} EXP(I_i, (m-1)c^{-1}).$$

The proof is then completed by checking the porosity index of each T_{mj} at each of its points. For this observe that, if x is a point of T_{mj}, then x is in C_m so that there are infinitely many intervals I_i for which

$$d(x, I_i) < mc^{-1} |I_i|.$$

But the intervals $EXP(I_i, (m-1)c^{-1})$ are complementary to T_{mj} and some elementary arithmetic shows that the porosity index of T_{mj} at x must then be at least c. This completes the proof.

We conclude this section by reporting on some other features of σ-porous sets that have been obtained in the literature. The first of these is due to Foran and Humke [78]; in this same paper a proof for the second lemma, but for F_σ σ-porous sets, is given. The version here, with G_δ σ-porous sets, is due to Tkadlec [240]. These results exhibit the rather strange containment properties of such sets.

($A_{4.3}$) LEMMA. Every σ-porous set is contained in a $G_{\delta\sigma}$ σ-porous set.

($A_{4.4}$) LEMMA. There is a porous set contained in no G_δ σ-porous set.

We have noted that the class of σ-porous sets is properly contained in the class of sets that are both first category and measure zero. One might therefore expect that certain well known properties of this latter class could be reduced further to the smaller class. Tkadlec [240] has shown the following, illustrating the special nature of the σ-porous sets in the larger classes.

($A_{4.5}$) LEMMA. There exists a perfect set P of measure zero, which is non σ-porous, such that the set $P + P$ contains no interval.

($A_{4.6}$) LEMMA. There exists an uncountable family of disjoint non σ-porous perfect sets.

§A5. Porosity lemmas.

Porosity requirements enter naturally into a number of questions of analysis. We can illustrate this with a remarkably simple and useful lemma due to M. J. Evans and P. D. Humke. It appears explicitly as a computational tool in their proof (Evans and Humke [73]) that the two upper (lower) Dini derivatives agree for a monotonic function except on a σ-porous set, but similar computations may be found throughout the literature.

(A5.1) EVANS-HUMKE POROSITY LEMMA. Let f be a nondecreasing function such that at a point x_0, $\underline{D}^+ f(x_0) < r < s$. Then the set of points
$$Y_s = \left\{ y : \frac{f(y) - f(x_0)}{y - x_0} > s \right\}$$
has porosity at least $1 - r/s$ on the right at x_0. Similarly if $\overline{D}^+ f(x_0) > s > r$ then the set of points
$$Y_r = \left\{ y : \frac{f(y) - f(x_0)}{y - x_0} < r \right\}$$
has porosity at least $1 - r/s$ on the right at x_0.

PROOF. Write $\theta = \frac{r}{s}$ so that $0 < \theta < 1$. Then we shall produce a sequence of intervals
$$I_k = (x_0 + \theta h_k, x_0 + h_k)$$
with the sequence $\{h_k\}$ decreasing to zero so that each interval I_k is disjoint from Y_s. From this it will follow that the porosity of Y_s on the right at x_0 must be at least $1 - \theta = 1 - \frac{r}{s}$ as required.

Note that $\underline{D}^+ f(x_0) < r$ so that there must be a sequence of points $x_0 + h_k$ for which $h_k \downarrow 0$ and
$$\frac{f(x_0 + h_k) - f(x_0)}{h_k} < r.$$
This sequence does the required task since if y is any point in an interval $(x_0 + \theta h_k, x_0 + h_k)$ then simply from the fact that f is nondecreasing we must have
$$\frac{f(y) - f(x_0)}{y - x_0} < \frac{f(x_0 + h_k) - f(x)}{h_k} \cdot \frac{h_k}{y - x_0} < \frac{r}{\theta} = s$$
so that no such point y can belong to Y_s as required.

For Y_r the computations are similar, but for completeness we give the details. If $\overline{D}^+ f(x_0) > s$ then we may select a sequence of points $\{x_0 + h_k\}$ with h_k converging down to zero and so that
$$\frac{f(x_0 + h_k) - f(x_0)}{h_k} > s.$$
Then again with $\theta = r/s$ we write
$$I_k = (x_0 + h_k, x_0 + \theta^{-1} h_k).$$
If y is any point inside an interval I_k we will have
$$\frac{f(y) - f(x_0)}{y - x_0} > \frac{f(x_0 + h_k) - f(x_0)}{h_k} \cdot \frac{h_k}{y - x_0}$$
and then since

$$y - x_0 < \theta^{-1} h_k \qquad (\text{i.e. } \frac{h_k}{y - x_0} > \theta)$$

we must have

$$\frac{f(y) - f(x_0)}{y - x_0} > s\theta = r.$$

This means that each interval I_k is disjoint from Y_r and again a simple porosity computation shows that the right hand porosity $p^+(Y_r; x_0)$ must exceed $1 - \theta = 1 - r/s$ as required. This completes the proof of the lemma.

The porosity lemma is sharp in that the estimates on the porosity cannot be improved. We give this in an example (from Bruckner and Thomson [44]).

(A5.2) Example. For any number $0 < p < 1$ take the sequence

$$h_n = (1 - p)^{2n}$$

and define the function f by setting

$$f(x) = (1 - p) h_{n-1} \qquad (h_n < x \leq h_{n-1})$$

and $f(x) = 0$ otherwise. Then one checks easily that $\underline{D}^+ f(0) = 1 - p$ and that the set Y,

$$Y = \{ y : \frac{f(y) - f(0)}{y} \geq 1 \}$$

is complementary to the intervals

$$((1 - p) h_{n-1}, h_{n-1}].$$

Thus Y has porosity p on the right at 0, which is just the estimate that the lemma provides.

Similar assertions, of course, can be obtained for nonincreasing functions and for porosity computations on either side. This lemma can be slightly generalized to apply to funtions that satisfy a given semi-Lipshitz condition. The computations are only mildly more complicated, but for completeness and reference we give the details.

(A5.3) LEMMA. Suppose that the function f satisfies the semi Lipshitz condition

$$\frac{\Delta f(I)}{|I|} > -M \qquad \text{for all intervals } I,$$

and that $\underline{D}^+ f(x_0) < r < s$. Then the set of points

$$Y = \left\{ y : \frac{f(y) - f(x_0)}{y - x_0} > s \right\}$$

has porosity on the right at least $1 - (M + r)/(M + s)$ at x_0.

Similarly if $\overline{D}^+ f(x_0) > s > r$ then the set

$$Y = \left\{ y : \frac{f(y) - f(x_0)}{y - x_0} < r \right\}$$

has porosity on the right at least $1 - (M + r)/(M + s)$.

PROOF. The proof is the same as that for the Evans-Humke lemma with only some changes in arithmetic. The details follow: set $\theta = (M + r)/(M + s)$ and note that the semi-Lipshitz condition assures that $-M < r < s$ so that θ lies in the interval $(0, 1)$. If $\underline{D}^+ f(x_0) < r$

then we may select a sequence of numbers $\{h_k\}$ decreasing to zero in such a way that
$$\frac{f(x_0+h_k) - f(x_0)}{h_k} < r .$$
We show that each interval
$$I_k = (x_0 + \theta h_k, x_0 + h_k)$$
is disjoint from the set Y,
$$Y = \{ y : \frac{f(y) - f(x_0)}{y - x_0} > s \} .$$
For if y belongs to an interval I_k then
$$\frac{f(y) - f(x_0)}{y - x_0}$$
$$= \frac{f(y) - f(x_0+h_k)}{y - x_0} + \frac{f(x_0+h_k) - f(x_0)}{y - x_0}$$
$$< \frac{M(x_0+h_k - y)}{y - x_0} + \frac{f(x_0+h_k) - f(x_0)}{h_k} \frac{h_k}{y - x_0}$$
$$< \frac{M(1-\theta) h_k}{\theta h_k} + \frac{r}{\theta}$$
$$< \theta^{-1}(M - M\theta + r) = \theta^{-1}(\theta M + \theta s - M\theta) = s .$$
It then follows by an easy porosity computation that $p^+(Y;x_0) > 1 - \theta$ and the first part of the lemma is proved.

For the second part if $\overline{D}^+ f(x_0) > s > r$ and Y denotes the set
$$Y = \left\{ y : \frac{f(y) - f(x_0)}{y - x_0} < r \right\}$$
then we select as before numbers h_k with h_k converging down to zero and with
$$\frac{f(x_0 + h_k) - f(x_0)}{h_k} > s .$$
Then again, if y is any point in an interval
$$I_k = (x_0 + \theta^{-1} h_k, x_0 + h_k) ,$$
we have
$$\frac{f(y) - f(x_0)}{y - x_0} >$$
$$\frac{f(y) - f(x_0+h_k)}{y - x_0} + \frac{f(x_0+h_k) - f(x_0)}{y - x_0}$$
$$> -\frac{M(y - [x_0 + h_k])}{y - x_0} + \frac{f(x_0+h_k) - f(x_0)}{h_k} \frac{h_k}{y - x_0}$$
$$> -M + (M+s) \frac{h_k}{y - x_0} > -M + (M+s) \theta = r .$$
Again this allows the porosity computation to be made and the lemma is proved.

There are a number of interesting, if elementary, consequences of these porosity computations that we present here as examples to illustrate the possible applications. More detailed applications appear in the main body of the text. (See especially chapter five.)

(A$_{5.4}$) **Example.** If f is monotonic and has a derivative $f_E'(x_0)$ at a point x_0 relative to a set E that is nonporous at x_0 then necessarily f is differentiable at x_0.

(A$_{5.5}$) **Example.** If f is monotonic and has a zero (resp. infinite) derivative $f_E'(x_0)$ at x_0 relative to a set E that is not strongly porous at x_0 then f has in fact a zero (infinite) derivative at x_0 in the ordinary sense.

These are easy consequences of the porosity lemma. It may be checked that the assertions here are sharp so that set porosity is precisely the right notion to capture these properties. Of course one-sided versions are also available. A more delicate version of these has some interest. See also section 65 in the text.

(A$_{5.6}$) **Example.** Let f be monotonic nondecreasing and suppose that $f_E'(x_0) = q$, i.e. that f has a derivative q at x_0 relative to a set E, where the porosity of E at x_0 does not exceed $p < 1$. Then

$$q(1-p) \leq \underline{D} f(x_0) \leq \overline{D} f(x_0) \leq \frac{q}{1-p}.$$

§A$_6$. Zahorski's porosity derivative.

The estimates in the preceding section regard the computation of the derivates for functions that satisfy some growth condition. In this section we discuss the converse direction by presenting an estimate of Zahorski [265] that shows how the increments of a function must behave when the derivates are given. Since the notion uses porosity computations we shall refer to it as a type of porosity derivative. It provides a generalization of the sharp derivative used in the text (see (7.7)).

(A$_{6.1}$) **DEFINITION.** Let $c \geq 0$ and let f be a real function. We define the lower, right, Zahorski derivate of f, $\underline{ZD}_c^+ f(x)$ with index $c > 0$ as,

$$\lim_{\varepsilon \to 0+} \inf \left\{ \frac{f(x+h+k) - f(x+h)}{k} : 0 \leq h,\ 0 < h+k < \varepsilon,\ \frac{k}{h} > c \right\}$$

and for the limiting case $c = 0$ we take the limit as $c \to 0+$.

The other three derivates $\overline{ZD}_c^+ f(x)$, $\underline{ZD}_c^- f(x)$, and $\overline{ZD}_c^- f(x)$ are similarly defined.

The basic result is just an elementary computation.

(A$_{6.2}$) **LEMMA.** Let $c > 0$. Then for any real function f we must have the inequalities

$$\overline{D}^+ f(x) \leq \overline{ZD}_c^+ f(x) \leq \overline{D}^+ f(x) + \frac{1}{c}\left(\overline{D}^+ f(x) - \underline{D}^+ f(x)\right),$$

and similarly for the three other Zahorski derivates.

PROOF. The proof follows directly from an entirely elementary computation:

$$\frac{f(x+h+k) - f(x+h)}{k} = \frac{f(x+h+k) - f(x)}{h+k} +$$
$$\frac{h}{k} \left[\frac{f(x+h+k) - f(x)}{h+k} - \frac{f(x+h) - f(x)}{h} \right].$$

($A_{6.3}$) COROLLARY. A real function f has a finite derivative on the right equal to d if and only if
$$\overline{ZD}_c{}^+ f(x) = \underline{ZD}_c{}^+ f(x) = d$$
for every index $c \geq 0$.

PROOF. This follows directly from the lemma.

($A_{6.4}$) COROLLARY. If a function f has a finite right hand derivative $f_+'(x)$ less than a number d and $\{I_n\}$ is any sequence of intervals converging to x on the right for which
$$\Delta f(I_n) \geq d |I_n|$$
for all n, then necessarily
$$\lim_{n \to \infty} \frac{|I_n|}{d(x, I_n)} = 0.$$

PROOF. This follows from the preceding corollary.

As further corollaries we may prove quite easily that a finite derivative has Zahorski's M_3-property, and that a bounded derivative has Zahorski's M_4-property.

($A_{6.5}$) COROLLARY. Let the function f have everywhere a finite derivative. Then for any number α each of the sets
$$A_\alpha = \{ x : f'(x) > \alpha \} \quad \text{and} \quad A^\alpha = \{ x : f'(x) < \alpha \}$$
is nonporous at each of its points.

PROOF. If $I = (a, b)$ is an interval disjoint from the set A^α then $f'(x) \geq \alpha$ everywhere on (a, b) and consequently, by an elementary monotonicity theorem,
$$\Delta f(I) = f(b) - f(a) \geq \alpha(b - a).$$
The preceding corollary now gives an estimate on the size of the interval I compared to the distance $d(x, I)$ if $f'(x) < \alpha$. The fact that A^α is nonporous at each of its points now follows.

For the M_4-property we introduce a type of generalized density parallel to the Zahorski derivates. For any $c > 0$ and any set A we write
$$\underline{ZD}_c{}^+ (A,x) = \liminf \frac{|A \cap (x+h, x+h+k)|^e}{k}$$
where lim is taken as $h, k \to 0+$ with $k/h > c$. Similarly the left lower density is defined for each c; the bilateral lower density is the minimum of the two unilateral densities and, finally, $\underline{ZD}_0(A,x)$ is the limit of this density as $c \to 0+$. The M_4-property concerns the density properties of the associated sets that appear in the preceding corollary.

(A$_{6.6}$) COROLLARY. Let the function f be Lipshitz and everywhere differentiable. Then for any number α each of the sets
$$A_\alpha = \{ x : f'(x) > \alpha \} \quad \text{and} \quad A^\alpha = \{ x : f'(x) < \alpha \}$$
may be expressed as a denumerable union of closed sets F_n such that
$$\inf{}_{x \in F_n} \underline{ZD}_\bullet (A, x) > 0 \qquad (A = A_\alpha \text{ or } A = A^\alpha).$$

PROOF. Since f' is in the first Baire class there is a sequence of closed sets F_n whose union is the set A_α and such that
$$\inf \{ f'(t) : t \in F_n \} = \alpha_n > \alpha.$$
We compute the density of $A = A_\alpha$ at a point x with $f'(x) > \alpha$. Let us write M for the Lipschitz constant (so that $|f'(t)| \leq M$) and $g(t) = f(t) - \alpha t$. Then, for $k/h > c$,

$$\frac{1}{k} | A \cap (x+h, x+h+k)| = \frac{1}{k} \int_{x+h}^{x+h+k} \chi_A(t)\, dt$$

$$\geq \frac{1}{k} \int_{x+h}^{x+h+k} \frac{g'(t)}{M}\, dt$$

$$= \frac{1}{M} \left[\frac{g(x+h+k) - g(x+h)}{k} \right] \to \frac{1}{M} \underline{Z D}_c\, g(x).$$

Now $g'(x) \geq \alpha_n - \alpha > 0$ for each $x \in F_n$ and so the estimate on the density for the set A follows from Lemma (A$_{6.2}$).

Finally let us conclude with a related porosity computation that is simple, but occasionally useful. This, too, concerns the density of a set taken in a porosity sense, but we shall express the computation in conventional language.

(A$_{6.7}$) LEMMA. Let $0 < \theta < 1$ and suppose that the set E has right, exterior, upper density greater than a positive number ρ. Then
$$\limsup{}_{h \to 0+} \frac{|E \cap (x + \theta h, x + h)|}{(1 - \theta)h} > \rho.$$

PROOF. If this were not the case then one would have
$$|E \cap (x + \theta h, x + h)| \leq (1 - \theta) h \rho$$
for all sufficiently small h. In particular, then for such h and every integer n,
$$|E \cap (x + \theta^n h, x + \theta^{n-1} h)| \leq (1 - \theta) \theta^{n-1} h \rho.$$
This gives
$$|E \cap (x, x + h)| \leq \sum_{n=1}^{\infty} |E \cap (x + \theta^n h, x + \theta^{n-1} h)|$$
$$\leq \sum_{n=1}^{\infty} (1 - \theta) \theta^{n-1} h \rho = h \rho.$$
But this contradicts the density requirement on E, and so the lemma is proved.

§A7. ψ-porosity.

The porosity measures the relative sizes of the gaps in a set E. A natural generalization of this is to measure this relative size in some other appropriate fashion. The simplest generalization is to take the same form of definition of ordinary porosity but to expand or contract the size of the gap by some appropriate function ψ. We define firstly the class of functions that we use and then give the definition of ψ-porosity index. It is more convenient to generalize the porosity index than to generalize the porosity itself and so we shall follow this course.

(A7.1) DEFINITION. A nonnegative function ψ defined on $[0, +\infty)$ is said to be an <u>admissable porosity function</u> provided, $\psi(0) = 0$, and ψ is increasing.

(A7.2) DEFINITION. Let ψ be an admissable porosity function. Then we define the (ψ)-<u>porosity index</u>, right and left, of a set E at a point x, by setting $PI_\psi^+(E; x)$ as,

$$\lim_{\epsilon \to 0+} \sup \left\{ \frac{\psi(k)}{h} : (x+h, x+h+k) \cap E = \emptyset, h+k < \epsilon, 0 \leq h, 0 \leq k \right\}$$

and $PI_\psi^-(E; x)$ as,

$$\lim_{\epsilon \to 0+} \sup \left\{ \frac{\psi(k)}{h} : (x-h-k, x-h) \cap E = \emptyset, h+k < \epsilon, 0 \leq h, 0 \leq k \right\}.$$

We adopt the same language as that introduced in section A_2 but employing the (ψ)-porosity index. Thus a set is non-(ψ)-porous at a point if its (ψ)-porosity index is zero, and is strongly (ψ)-porous if its (ψ)-porosity index is $+\infty$. A set is (ψ)-porous if each point is a point of (ψ)-porosity index greater than zero, and a set is σ-(ψ)-porous if it may be expressed as a union of a sequence of sets each of which is (ψ)-porous.

The usual choices of porosity function ψ are as follows.

(A7.3) Taking $\psi(t) = t$ we have the ordinary porosity index defined earlier. For convenience we shall call this ordinary porosity or (x)-porosity.

(A7.4) Taking $\psi(t) = t^\alpha$ for $0 < \alpha < 1$ we have a type of porosity first studied by Yanagihara [256] and Zajíček [269]. Following the notation of the latter of these authors we shall refer to this as (x^α)-porosity. Note that we use the (ψ)-porosity index rather than the (ψ)-porosity as computed from a generalization of definition ($A_{2.2}$), i.e. by writing

$$p_\psi^+(E; x) = \lim \sup_{h \to 0+} \frac{\psi(\lambda(E, x, x+h))}{h}.$$

In the special case $\psi(t) = t^\alpha$ one may check that the (x^α)-porosity and the (x^α)-porosity index are identical for $0 < \alpha < 1$ (but not of course for $\alpha = 1$).

(A7.5) Taking $\psi(t) = t^\beta$ for $\beta > 1$ gives a broader type of porosity that has been used by Haussermann [110] (with a somewhat different formulation) in his study of the behaviour of typical continuous functions.

(A7.6) Finally taking $\psi(0) = 0$ and $\psi(t) = 1+t$ for $t > 0$ we obtain a limiting case in which (ψ)-porous sets are merely nowhere dense, and σ-(ψ)-porous sets are first category.

In order to provide some substance and insight into the nature of sets having various porosity requirements we consider some examples. A sequence $\{h_n\}$ converging to zero is said to have a porosity p in one of these senses (right porosity, left porosity, right (x^α)-porosity, etc.) if the set of points
$$\{h_1, h_2, h_3, h_4, \ldots\}$$
has that porosity at the point 0. If $\{h_n\}$ is a descending sequence with $\lim h_n = 0$ and if
$$\limsup_{n \to \infty} \frac{h_n}{h_{n+1}} = r \quad (1 \leq r \leq +\infty)$$
then the sequence has right porosity index $r - 1$. Thus slowly converging sequences have zero porosity index. In the case of (x^α)-porosity the number s $(0 \leq s \leq +\infty)$,
$$\limsup_{n \to \infty} \frac{(h_n - h_{n+1})^\alpha}{h_{n+1}} = s,$$
is the right (x^α)-porosity index of the sequence $\{h_n\}$. If a sequence has right (x^α)-porosity finite ($s < +\infty$) for some $0 < \alpha < 1$ then the sequence has right porosity 0 in the ordinary sense, and also too in the (x^β)-sense for any $\alpha < \beta \leq 1$. Thus again, while for zero porosity the sequence must be slowly converging to zero, for the (x^α)-porosity to be finite it must be even more slowly converging to zero. These concepts allow a precise language for "slowly converging to zero" together with a tight interrelationship between these notions and various estimates for the Dini derivatives of various classes of functions.

We provide an example (from Bruckner and Thomson [44]) to show how one may generate sequences that exhibit certain (x^α)-porosity behaviour.

(A7.7) Example. Suppose that ψ is a continuous strictly increasing function on $[0, +\infty)$. We suppose that for an integer k,
$$\frac{\psi(t)}{t^s} \to 0 \quad \text{as} \quad t \to 0+.$$
where $s = 1/(k+1)$. We show how a sequence of numbers $\{x_n\}$ may be constructed that has zero (ψ)-porosity index on the right.

Let $\alpha(x) = x - x^{k+1}$ and for any $x \in (0,1)$ define the sequence $\{x_n\}$ by writing inductively
$$x_1 = \alpha(x), \, x_2 = \alpha(x_1), \, \ldots \, x_j = \alpha(x_{j-1}).$$
We claim that this sequence must have zero (ψ)-porosity index on the right. The sequence $\{x_n\}$ is decreasing to zero and so the porosity computation follows if we establish the limit
$$\lim_{t \to 0+} \frac{\psi(t - \alpha(t))}{t} = 0.$$
But
$$\frac{\psi(t - \alpha(t))}{t} = \frac{\psi(t^{k+1})}{t}$$

and for t sufficiently close to zero $\psi^{-1}(t) > t^{k+^{1}/^{2}}$ so that $\psi(t) < t^{1/(k+^{1}/^{2})}$. This gives for small t,

$$\frac{\psi(t-\alpha(t))}{t} = \frac{\psi(t^{k+1})}{t} < t^{(k+1)/(k+^{1}/^{2}) - 1} = t^{1/(2k+1)}$$

and the limit is established as required.

In order to provide some feeling for the nature of such sequences we carry through the necessary computations in order to exhibit a sequence of numbers which has zero $(x^{1/2})$-porosity index but positive $(x^{1/3})$-porosity index; the numbers help illustrate what "slowly converging to zero" signifies. The sequence $\{x_n\}$ is computed as above using the function $\alpha(x) = x - x^3$.

n	x_n	x_n	x_n	x_n
1	.100000000000	.010000000000	.001000000000	.000100000000
2	.099000000069	.009999000000	.000999999000	.000099999999
3	.098029701062	.009998000300	.000999998000	.000099999998
4	.097087653028	.009997000900	.000999997000	.000099999997
5	.096172503685	.009996001800	.000999996000	.000099999996

(A7.8) Example. In the other direction here are some examples of sequences that are porous in coarser senses. The sequence $x_n = r^n$ for some number $0 < r < 1$ has infinite (x^α)-porosity index for $0 < \alpha < 1$, has index $(1-r)/r$ if $\alpha = 1$ and index 0 if $\alpha > 1$.

The sequence $x_n = (n!)^{-1}$ is strongly porous in the ordinary sense, but it is (x^α)-nonporous for every $\alpha > 1$.

The sequence $x_n = \exp(-\beta^n)$ $(\beta > 1)$ is (x^α)-strongly porous if $\alpha < \beta$, and (x^α)-nonporous if $\alpha > \beta$.

Let us conclude this section by observing that a number of the results of earlier sections for ordinary porosity may be readily extended to (ψ)-porosity. Thus we have the following analogues from section A_2, and we require no modifications in the proofs.

(A7.9) LEMMA. Let P be a perfect nowhere dense set and let $\{(a_n, b_n)\}$ be the sequence of intervals complementary to P. Then at any point $x \in P$ and not isolated on the right in P, and for a porosity function ψ

$$PI_\psi^+(P; x) = \limsup \frac{\psi(b_n - a_n)}{a_n - x}$$

as a_n, b_n converge to x, with $a_n > x$.

(A7.10) LEMMA. Let P be a perfect set and I_n the sequence of open intervals complementary to P. Then if x is a point of P that is not isolated on the right in P and which has right (ψ)-porosity index exceeding a positive number i then x must belong to infinitely many intervals

$$EXP(I_n, \frac{1}{i} \psi) \quad n = 1, 2, 3, \ldots.$$

(A7.11) LEMMA. Let ψ be any porosity function, let E be an arbitrary set and let i be a finite positive number. Then the set of points x at which the right [left, bilateral] (ψ)-porosity index of E at x is greater than i is a a set of type G_δ.

(A7.12) LEMMA. Let ψ be any porosity function. If P is perfect and nowhere dense then the set of points in P at which P is bilaterally strongly (ψ)-porous is residual in P.

§A8. σ-(ψ)-porosity.

There are a number of interrelations between σ-porosity taken in two different senses. In this section we explore some of these relations. The first is more or less immediate.

(A8.1) LEMMA. Let ψ_1 and ψ_2 be porosity functions for which
$$\lim_{t \to 0+} \frac{\psi_1(t)}{\psi_2(t)} = 0.$$
Then any set that is σ-(ψ_1)-porous is necessarily strongly ψ_2-porous.

PROOF. If a set E has ψ_1-porosity index on the right greater than a positive number p then there is a sequence of intervals $\{I_n\}$ converging to x on the right, each interval I_n is complementary to E, and
$$\frac{\psi_1(I_n)}{d(x, I_n)} > p.$$
This means that
$$\frac{\psi_2(I_n)}{d(x, I_n)} = \frac{\psi_2(I_n)}{\psi_1(I_n)} \frac{\psi_1(I_n)}{d(x, I_n)} \to +\infty,$$
and, consequently, E is strongly ψ_2-porous at x on the right. From this observation, together with its left hand counterpart, the lemma may be obtained.

In particular note that any set that is σ-porous in the ordinary sense must be strongly porous for any porosity function $\psi(t) = t^\alpha$ ($\alpha < 1$). Any set that is σ-(x^β)-porous for some $\beta > 1$ must be strongly porous in the ordinary sense.

A deeper relation has been obtained by Zajíček [269]. He has noted that while porosity in the sense (x^{α_1}) and in the sense (x^{α_2}) differ if $\alpha_1 \neq \alpha_2$ there is no distinction between σ-porosity taken in the two senses for $0 < \alpha_1 \leq \alpha_2 < 1$. This follows as a particular case of this general observation.

(A8.2) LEMMA. [Zajíček] Let ψ_1 and ψ_2 be porosity functions with the following property: for any $p > 0$ there are $\delta > 0$, $q > 0$, and an integer r so that the r-fold composition
$$(q\psi_1) \circ (q\psi_1) \circ (q\psi_1) \cdots (q\psi_1)(x) \geq p\, \psi_2(x) \qquad \text{for } 0 < x < \delta.$$
Then any set A that is σ-(ψ_2)-porous is necessarily also σ-(ψ_1)-porous.

PROOF. It is enough to prove that a bounded set A, that has ψ_2-porosity index greater than a positive number p on the right, may be expressed as a countable union of sets each of which has ψ_1-porosity index greater than q on the right, where q is as in the assertion of the lemma for p. Without loss of generality we may suppose that no point of A is isolated on the right, since there are denumerably many such points.

Let I_i, $i = 1, 2, 3, \ldots$ be the sequence of intervals complementary to A and let $\delta, q,$ and r be as in the assertion of the lemma with this p. Choose N so that $|I_i| < \delta$ if $i \geq N$. Let us write the sets

$$C_k = A \cap \left(\prod_{n=N}^{\infty} \sum_{i=n}^{\infty} \text{EXP}(I_i, (q\psi)^{(k)}) \right)$$

for $k = 1, 2, \ldots r$ where $(q\psi)^{(k)}$ is the k-fold composition

$$(q\psi_1) \circ (q\psi_1) \circ (q\psi_1) \cdots (q\psi_1)(x) \ .$$

Since the ψ_2-porosity index on the right at any point x in A exceeds p and since $(q\psi_1)^{(r)}(x) > p\,\psi_2(x)$ for $0 < x < \delta$, we see, using lemma (A7.10), that $C_r = A$. Set $C_0 = \emptyset$; we show that each of the sets

$$C_1 \setminus C_0, \ C_2 \setminus C_1, \ C_3 \setminus C_2, \ \ldots, \ C_r \setminus C_{r-1}$$

is σ-(ψ_1)-porous, and the lemma is then proved.

Define the sets T_{km}, by

$$T_{km} = C_k \setminus \left(\sum_{i=m}^{\infty} \text{EXP}(I_i, (q\psi_1)^{(k-1)}) \right)$$

(where for convenience $(q\psi)^{(0)} = 0$) and observe that

$$C_k = \sum_{m=1}^{\infty} T_{km} .$$

We claim that each set T_{km} is (ψ_1)-porous, indeed that at each point $x \in T_{km}$ the right (ψ_1)-porosity index of T_{km} exceeds q. This is a consequence of the fact that, by the nature of the construction x must be in infinitely many of the intervals

$$\text{EXP}\left(I_i, (q\psi_1)^{(k)}\right)$$

while each of the intervals

$$\text{EXP}\left(I_i, (q\psi_1)^{(k-1)}\right) \quad i = m, m+1, \ldots$$

is complementary to T_{km}. This exhibits each C_k as σ-(ψ_1)-porous and the lemma is proved.

As an immediate corollary we have the equality of the notions of σ-(x^{α_1})-porous sets and σ-(x^{α_2})-porous sets for $\alpha_1, \alpha_2 \in (0, 1)$. We have only to take $\psi_1(t) = t^{\alpha_1}$, $\psi_2(t) = t^{\alpha_2}$ for $\alpha_1 > \alpha_2$ and observe that the conditions of the lemma are obtained (with $q = 1$) since

$$(\psi_1) \circ (\psi_1) \circ (\psi_1) \circ \cdots (\psi_1)(x) = x^{\alpha_1^r} \geq p\,x^{\alpha_2} = p\,\psi_2(x)$$

for sufficiently small x and sufficiently large r. Thus we have the following corollary.

(A8.3) COROLLARY. For any α_1, α_2 in $(0,1)$ a set is σ-(x^{α_1})-porous if and only if it is σ-(x^{α_2})-porous.

These results are limited by the following observations. Proofs appear in Zajíček [269] and section A_{11} below.

(A8.4) LEMMA. For any $0 < \alpha < 1$ there exists a perfect set of measure zero which is (x^α)-porous but not σ-porous in the ordinary sense.

(A8.5) LEMMA. For any $0 < \alpha < 1$ there is a perfect set of positive Lebesgue measure which is (x^α)-porous.

§A9. Porosity and Hausdorff measure.

There is a connection between the thinness of a set estimated in the porosity sense and its thinness given in the more familar sense of Hausdorff dimension. In this section we prove some of the interrelations that are possible. This material is due to Haussermann [110].

Let h be a continuous strictly increasing positive function on $(0, +\infty)$ and let μ_ϵ^h ($\epsilon > 0$) denote the outer measure defined by writing

$$\mu_\epsilon^h(X) = \inf \sum_{i=1}^\infty h(|I_i|)$$

where the infimum is taken with regard to all sequences of intervals $\{I_n\}$ that cover the set X and for which $|I_n| < \epsilon$. Then we write

$$\mu^h = \lim_{\epsilon \to 0+} \mu_\epsilon^h.$$

This latter measure is called the h-Hausdorff measure. The choices $h(x) = x^\alpha$ ($0 < \alpha < 1$) are the most familiar. The connection we establish is between (ψ)-porosity and h-Hausdorff measure where ψ and h are functions inverse to each other.

(A9.1) LEMMA. Let ψ be a continuous, strictly increasing porosity function and h its inverse. If A is the set of points at which the right (ψ)-porosity index of a set E exceeds 1 then $\mu^h(A) = 0$.

PROOF. We shall use the notation $ADJ^+(I, \psi)$ to indicate that interval adjacent to I on the right and with length $\psi(|I|)$; this is closely related to the notation $EXP(I, \psi)$. For the proof we may assume that E is bounded and ignore the denumerable collection of points in A that are isolated on the right in E. If $\{I_n\}$ denotes the collection of open intervals complementary to the set E, we know that every point x in A must belong to infinitely many intervals,

$$ADJ^+(I_n, \psi) = J_n.$$

Note that $h(J_n) = h(\psi(|I_n|)) = |I_n|$.

Let $\varepsilon > 0$. We choose an integer n_0 so large that

$$\sum_{n=n_0}^{\infty} |I_n| < \varepsilon \quad \text{and} \quad \psi(|I_n|) < \varepsilon \quad \text{if} \quad n \geq n_0.$$

The collection of intervals $\{J_n : n \geq n_0\}$ covers the set A (i.e. it covers the set of points that are not isolated on the right in E), each interval has length less than ε and the sum of the $h(|J_n|)$ is less than ε. This gives

$$\mu_\varepsilon^h(A) \leq \sum_{n=n_0}^{\infty} h(|J_n|) < \varepsilon.$$

Since $\varepsilon > 0$ is arbitrary it follows that $\mu^h(A) = 0$ as required.

(A9.2) LEMMA. Let $0 < \alpha < 1$, $1 < \beta < +\infty$ with $\beta = \frac{1}{\alpha}$. If a set E is (x^β)-σ-porous then it must have (x^α)-Hausdorff measure zero, and so in particular E has Hausdorff dimension no larger than α.

PROOF. It is enough to suppose that E is itself (x^β)-porous, and indeed we may suppose that E has right (x^β)-porosity index exceeding $c > 0$ at each point. Define the functions ψ and h by

$$\psi(t) = \frac{t^\beta}{c} \quad \text{and} \quad h(t) = (ct)^\alpha$$

and apply lemma (A9.1) above. We must have that $\mu^h(E) = 0$. From this we may check that the (x^α)-Hausdorff measure of E also vanishes and the lemma is proved.

§A10. A further porosity lemma.

Our porosity lemma (A5.1) permits a generalization to the class $C(\psi)$ of continuous functions f that satisfy an inequality of the form

$$|f(x) - f(y)| \leq \psi(|x - y|) \qquad (|x - y| \leq 1)$$

where ψ is a given modulus of continuity, i.e. ψ is defined for all nonnegative reals, is increasing, and

$$\lim_{x \to 0+} \psi(x) = \psi(0) = 0.$$

Of course the most interesting special case occurs with $\psi(x) = M x^\alpha$ for $0 < \alpha \leq 1$, and for M a positive real constant.

(A10.1) LEMMA. Let f be a continuous function that satisfies an inequality

$$|f(x) - f(y)| \leq \psi(|x - y|) \quad (|x - y| \leq 1)$$

for a modulus of continuity ψ that is defined and continuous on $[0, +\infty)$, with $\psi(0) = 0$ and with $\psi_+'(0) = +\infty$. Suppose that at a point x_0 one has $\underline{D}^+ f(x_0) < r < s$. Then the set of points

$$Y = \left\{ y : \frac{f(y) - f(x_0)}{y - x_0} \geq s \right\}$$

has right (ψ)-porosity at x_0 at least $s - r$. Similarly if $\overline{D}^+ f(x_0) > s > r$ then the set of

points
$$Y = \{y : \frac{f(y) - f(x_0)}{y - x_0} \leq r\}$$
has right (ψ)-porosity at x_0 at least $s - r$.

PROOF. We prove the first statement of the lemma. Since $\underline{D}^+ f(x_0) < r$ we may select a sequence of positive numbers $\{h_n\}$ descending to zero so that
$$\frac{f(x_0 + h_n) - f(x_0)}{h_n} < r.$$
We will show how to choose numbers $\{\theta_k\}, 0 < \theta_k < 1$, and a subsequence $\{h_{n_k}\}$ in such a way that $\theta_k \to 1$,
$$\frac{1}{\theta_k}\left[\frac{\psi(h_{n_k}(1-\theta_k))}{h_{n_k}} + r\right] = s.$$
and
$$\lim_{k \to \infty} \frac{\psi((1-\theta_k)h_{n_k})}{h_{n_k}} = s - r.$$

Let us suppose for the moment that these numbers can be so chosen. If so then we can obtain the porosity estimate promised in the statement of the lemma. Consider the set of points
$$Y = \{y : \frac{f(y) - f(x_0)}{y - x_0} \geq s\}$$
and the sequence of intervals
$$\{(x_0 + \theta_k h_{n_k}, x_0 + h_{n_k})\}.$$
We observe that each of these intervals is necessarily disjoint from Y since if there is given a point y,
$$x_0 + \theta_k h_{n_k} < y < x_0 + h_{n_k},$$
we must have using the ψ-inequality on f that
$$\frac{f(y) - f(x_0)}{y - x_0} =$$
$$\frac{f(y) - f(x_0 + h_{n_k})}{y - x_0} + \frac{f(x_0 + h_{n_k}) - f(x_0)}{h_{n_k}} \cdot \frac{h_{n_k}}{y - x_0}$$
$$< \frac{\psi(y - x_0 - h_{n_k})}{y - x_0} + \frac{r}{\theta_k}$$
$$< \frac{1}{\theta_k}\left[\frac{\psi((1-\theta_k)h_{n_k})}{h_{n_k}} + r\right] = s.$$

As each interval $(x_0 + \theta_k h_{n_k}, x_0 + h_k)$ is now seen to be disjoint from the set Y we compute that the (ψ)-porosity of Y on the right must exceed the number
$$\limsup_{n \to \infty} \frac{\psi((1-\theta_k)h_{n_k})}{h_{n_k}}.$$
But from the way in which these sequences have been chosen this limit is $s - r$ which is exactly the porosity which we were required to obtain.

Thus it remains for us to prove that these sequences may be selected in the way that we have stated. Consider in the (ξ, y) plane for $0 \leq \xi \leq 1$ the straight line $y = s\xi - r$ and for any $h > 0$, the curve

$$y = \frac{1}{h} \psi((1-\xi)h).$$

For fixed $\xi < 1$ the limit

$$\lim_{h \to 0+} \frac{\psi((1-\xi)h)}{h} = (1-\xi)\psi_+'(0) = +\infty.$$

Note that the line passes through the point $(1, s-r)$, and the curve passes through the point $(1, 0)$. This allows us to select points $\xi_1, \xi_2, \xi_3, \ldots$ and indices $n_1, n_2, n_3, \ldots$ inductively so that $n_1 = 1$, and

$$s\xi_1 - r > \frac{\psi((1-\xi_1)h_1)}{h_1},$$

$$s\xi_1 - r < \frac{\psi((1-\xi_1)h_{n_2})}{h_{n_2}},$$

$$\xi_2 > 1 - 1/2,$$

$$s\xi_2 - r > \frac{\psi((1-\xi_2)h_{n_2})}{h_{n_2}},$$

and so on in this fashion so that $\{\xi_j\}$ and $\{h_{n_j}\}$ are obtained and satisfy the inequalitites

$$\xi_j > 1 - \frac{1}{j}$$

$$s\xi_j - r < \frac{\psi((1-\xi_j)h_{n_{j+1}})}{h_{n_{j+1}}}$$

$$s\xi_{j+1} - r > \frac{\psi((1-\xi_{j+1})h_{n_{j+1}})}{h_{n_{j+1}}}.$$

Since ψ is continuous we may choose numbers $\{\theta_j\}$ from the intervals (ξ_j, ξ_{j+1}) so that

$$s\theta_j - r = \frac{\psi((1-\theta_j)h_{n_{j+1}})}{h_{n_{j+1}}}.$$

Since these sequences are precisely what was required, the proof is complete.

Note that, in the lemma which we have just proved, the extreme case with $\psi(t) = +\infty$ ($t > 0$) and $\psi(0) = 0$ is not permitted in the assertion. Nonetheless the lemma is true for such a function ψ since, in that case, positive (ψ)-porosity of a set A at a point is equivalent to the nondenseness of the set A at that point.

§A_{11}. Symmetric perfect sets.

A number of examples of porosity computations in the setting of symmetric perfect sets help illuminate the ideas. For the definition and basic properties of such sets see Bari [9]. We review the construction here.

Let $\xi = \{\xi_n\}$ be a sequence of real numbers from the interval $(0, 1/2)$. We define the following subsets of the interval $[0, 1]$:

$$J_1^1 = [0, \xi_1], \quad J_1^2 = [1-\xi_1, 1], \quad I_1^1 = (\xi_1, 1-\xi_1)$$

$$C_1 = J_1^1 \cup J_1^2,$$

$$J_2{}^1 = [0, \xi_1\xi_2], \quad J_2{}^2 = [\xi_1 - \xi_1\xi_2, \xi_1],$$
$$J_2{}^3 = [1-\xi_1, 1-\xi_1+\xi_1\xi_2], \quad J_2{}^4 = [1-\xi_1\xi_2, 1]$$
$$I_2{}^1 = (\xi_1\xi_2, \xi_1-\xi_1\xi_2), \quad I_2{}^2 = (1-\xi_1+\xi_1\xi_2, 1-\xi_1\xi_2),$$
$$C_2 = J_2{}^1 \cup J_2{}^2 \cup J_2{}^3 \cup J_2{}^4,$$

and so on. Thus each $J_n{}^k$ is a closed interval of length $\xi_1\xi_2\cdots\xi_n$ with k ranging from 1 to 2^n, the intervals $I_n{}^i$ ($1 \leq i \leq 2^{n-1}$) are open and complementary to the J's of the same stage. The sets

$$C_n = J_n{}^1 \cup J_n{}^2 \cup \ldots J_n{}^{2^n}$$

form a decreasing sequence of closed sets, and finally the set

$$C(\xi) = \prod_{n=1}^{\infty} C_n$$

is a nowhere dense perfect subset of the interval $[0, 1]$, that has measure

$$|C(\xi)| = \lim_{n \to \infty} 2^n \xi_1 \xi_2 \xi_3 \cdots \xi_n.$$

We call this the **symmetric perfect set** generated by the sequence ξ. In the special case where each $\xi_n = 1/3$ the set $C(\xi)$ is exactly the usual Cantor ternary set. Note the construction is very close to the Cantor construction but rather than removing the "middle thirds", intervals of proportion $\alpha_n = (1-2\xi_n)$ are removed at the nth stage.

This is the classical notation for these sets. The reader may prefer to rewrite this construction and reformulate the proof in more modern notation. We would write Σ for the collection of all sequences of 0's and 1's and Σ_n for the collection of all length n sequences of 0's and 1's. Inductively we define J_σ and I_σ for $\sigma \in \Sigma_n$ by writing

(i) $J_\emptyset = [0, 1]$,

(ii) for $n \geq 0$ and $\sigma \in \Sigma_n$,

$$EXP(I_\sigma, \frac{\xi_n}{\alpha_n}) = J_\sigma,$$

(iii) for $n \geq 0$ and $\sigma \in \Sigma_n$, $J_{\sigma 0}$ and $J_{\sigma 1}$ are the left and right components of the set

$$J_\sigma \setminus int\, I_\sigma.$$

Then the set $C(\xi)$ is defined as

$$C(\xi) = \prod_{n=0}^{\infty} \sum_{\sigma \in \Sigma_n} J_\sigma.$$

Our main theorem in this section is obtained as a series of porosity computations for such symmetric perfect sets; this appears in Humke and Thomson [122]. Following the proof of the theorem we illustrate its applications by providing a number of symmetric perfect sets having certain specified porosity characteristics.

(A11.1) THEOREM. Let $\xi = \{\xi_n\}$ be a sequence of numbers from the interval $(0, 1/2)$ and let $C = C(\xi)$ be the symmetric perfect set constructed from the sequence ξ. Then the following hold:

(i) If $\xi_n \to 1/2$ then C is not σ-porous.

(ii) If $\liminf \xi_n < 1/2$ then C is porous.

(iii) If $\liminf \xi_n = 0$ then C is strongly porous.

(iv) If $\liminf \xi_n > 0$ then C is not σ-strongly porous.

(v) If $0 < p < 1$ and
$$\limsup\, (1 - 2\xi_n)[\xi_1 \xi_2 \xi_3 \cdots \xi_n]^{1-p^{-1}} > 0$$
then C is (x^p)-porous at each of its points.

(vi) If
$$\lim \frac{1 - 2\xi_n}{[\xi_1 \xi_2 \xi_3 \cdots \xi_n]^s} = 0$$
for all $s \geq 0$ then C is not σ-(x^p)-porous for any $0 < p < 1$.

(vii) If $1 < p < +\infty$ and
$$\limsup \frac{[\xi_1 \xi_2 \xi_3 \cdots \xi_{n-1}]^{p-1}}{\xi_n} > 0$$
then C is (x^p)-porous at each of its points.

(viii) If $1 < p < +\infty$ and
$$\lim \frac{[\xi_1 \xi_2 \xi_3 \cdots \xi_{n-1}]^{p-1}}{\xi_n} = 0$$
then C is not σ-(x^p)-porous.

PROOF. Let us prove firstly the assertions (ii), (iii), and (v) which assert that the set C has some specified porosity. For the case (ii) let x be any point in the set C. Then at each stage n there is an interval J_n^i to which x belongs, and in that interval there is an interval $I(n, x) = I_{n+1}^j$ which is complementary to C. The distance from x to that latter interval cannot exceed
$$\xi_1 \xi_2 \xi_3 \cdots \xi_n \xi_{n+1}$$
while the length of that interval $I(n, x)$ itself is
$$(1 - 2\xi_{n+1})\xi_1 \xi_2 \xi_3 \cdots \xi_n .$$
This gives, for the ordinary porosity, the estimate
$$\frac{|I(n, x)|}{d(x, I(n, x))} \geq \frac{1 - 2\xi_{n+1}}{\xi_{n+1}} = \frac{1}{\xi_{n+1}} - 2 .$$
Certainly if the sequence ξ has a limit inferior less than $1/2$ then the ordinary porosity index of C at x must be positive; since x is an arbitrary point of C, the set C is porous so that (ii) is proved. If the limit inferior is zero then the porosity index must be $+\infty$ and this proves assertion (iii).

Similarly if one computes the (x^p)-porosity index in the same manner assertion (v) and (vii) will follow. For example we have

$$\frac{|I(n,x)|^p}{d(x, I(n,x))} \geq$$

$$\left[(1-2\xi_{n+1})\xi_1\xi_2\xi_3\cdots\xi_n\right]^p \left[\xi_1\xi_2\xi_3\cdots\xi_n\xi_{n+1}\right]^{-1}$$

$$= \left[(1-2\xi_{n+1})\left[\xi_1\xi_2\xi_3\cdots\xi_n\right]^{1-1/p}\right]^p \xi_{n+1}^{-1}.$$

If ξ_n does not converge to $1/2$ then C is (x^q)-porous for all $0 < q \leq 1$. If ξ_n does converge to $1/2$ and the lim sup in (iv) is positive then the above estimate gives again positive (x^p)-porosity index.

It remains to prove the more difficult assertions that, under the asserted conditions, the set fails to be σ-porous in certain senses. Let us prove assertion (i), namely that, under the assumption that $\xi_n \nrightarrow 1/2$, the set C is not σ-porous. In order to obtain a contradiction let us suppose that this set C is σ-porous so that it may, by ($A_{4.2}$), be written as

$$C = \{x_1, x_2, x_3, \ldots\} \cup \sum_{n=1}^{\infty} E_n$$

where each set E_n contains no isolated points and has porosity index exceeding 1 at each point. We shall define a sequence of compact sets $\{F_n\}$ in such a way that for each index n,

$$\emptyset \neq F_{n+1} \subset F_n \subset C, \quad x_n \notin F_n \quad \text{and} \quad E_n \cap F_n = \emptyset.$$

From this we obtain our contradiction since there must be a point $x \in C$ in the intersection of the F_n. But such a point does not appear in the sequence $\{x_n\}$, belongs to no E_n, and consequently cannot belong to C.

The construction is given inductively. We shall for each positive integer n define an integer $k(n)$, an interval $J(n)$, and a compact set F_n. Since $\xi_n \nrightarrow 1/2$ we may choose an integer $k'(n)$ so that for $k \geq k'(n)$,

$$|1 - 2\xi_k| < 3^{-n-2} \tag{1}.$$

For the first step ($n=1$) we consider two possibilities. In case E_1 is not dense in C, there must be a first integer $k(1) > k'(1)$ such that one of the intervals $J_{k(1)}{}^i$ contains no point of $E_1 \cup \{x_1\}$. This defines $k(1)$ and we take $J(1)$ as that interval.

On the other hand if E_1 is dense in C then we take $k(1) = k'(1)$, and $J(1)$ as any interval $J_{k(1)}{}^i$ that does not contain x_1. In both cases we define the set F_1 to be the closed subset of C,

$$J(1) \setminus \sum_{n=k(1)+1}^{\infty} \sum \left\{ \text{EXP}(I_n{}^i, 1) : 1 \leq i \leq 2^{n-1}, I_n{}^i \subset J(1) \right\} \quad (2)$$

obtained by expanding the intervals complementary to C in $J(1)$. Note that F_1 is a nonempty closed subset of C. Certainly F_1 is closed and it is contained in C since it is the complementary intervals to C that are expanded and removed. That it is nonempty follows from the requirement (1) that the intervals removed are suffciently small. We check also that F_1 contains no point of E_1. In the situation in which E_1 is not dense in C this is immediate; in the other sitatuation we see that F_1 can contain no point of E_1 because of the porosity requirements. The intervals in (2) that are expanded and removed from $J(1)$ are exactly the complementary intervals to E_1; since every point of E_1 has porosity index at least 1 these points appear in the set that is removed from $J(1)$.

We continue inductively. For convenience let us introduce the notation $\text{EXP}^m(I, 1)$ ($m = 1, 2, \ldots$) so that

$$\text{EXP}^{m+1}(I, 1) = \text{EXP}\left(\text{EXP}^m(I, 1), 1\right).$$

The notation then indicates that the interval I has been successively expanded, m times. Note that

$$\left| \text{EXP}^m(I, 1) \right| = 3^m |I|$$

Suppose now that we have defined $k(n)$, $J(n)$, and F_n for $n = 1, 2, \ldots, m$ and let us define $k(m+1)$, $J(m+1)$, and F_{m+1}. If E_{m+1} is not dense in F_m then there is a first integer $k(m+1)$ with

$$k(m+1) > \max\{k'(m+1), k(m)\}$$

so that some component interval of the set

$$J(m) \setminus \sum_{n=k(m)+1}^{k(m+1)} \sum \left\{ \text{EXP}^m(I_n{}^i, 1) : 1 \leq i \leq 2^{n-1}, I_n{}^i \subset J(m) \right\} \quad (3)$$

is disjoint from $E_{m+1} \cup \{x_{m+1}\}$. We take $J(m+1)$ to be one such interval.

If, instead, E_{m+1} is dense in F_m then we take

$$k(m+1) = \max\{k'(m+1), k(m)\} + 1$$

and $J(m+1)$ is any component interval of the set (3) above that does not contain the point x_{m+1}. In both cases we define the set F_{m+1} as

$$J(m+1) \setminus \sum_{n=k(m+1)+1}^{\infty} \sum \left\{ \text{EXP}^{m+1}(I_n{}^i, 1) : 1 \leq i \leq 2^{n-1}, I_n{}^i \subset J(m+1) \right\} \quad (4)$$

By this procedure we define $\{F_n\}$ inductively and we may check the following facts. Each $F_n \neq \emptyset$. For each n, $F_{n+1} \subset F_n$. Each set F_n is disjoint from E_n. The arguments above for the case $n = 1$ may be essentially repeated here. Since this situation leads to a contradiction we have proved (i).

Assertion (iv) may be proved in almost the same way but with some changes in the arithmetic. If ξ_n is bounded above away from 0 then there is a positive number ε so that always $\xi_n > \varepsilon > 0$. Suppose that $C(\xi)$ is in this situation σ-strongly porous. Then there is a sequence of sets E_n as above whose union is C but each E_n has porosity index $+\infty$.

We repeat the above induction and essentially the same arguments but replace (2), (3), and (4) by (5), (6), and (7) respectively:

$$J(1) \setminus \sum_{n=k(1)+1}^{\infty} \sum_{i=1}^{2^n-1} EXP(I_n^i, \varepsilon) \tag{5}$$

$$J(m) \setminus \sum_{n=k(m)+1}^{k(m+1)} \sum_{i=1}^{2^n-1} E_m(I_n^i) \tag{6}$$

$$J(m+1) \setminus \sum_{n=k(m+1)}^{\infty} \sum_{i=1}^{2^n-1} E_{m+1}(I_n^i) \tag{7}$$

where, inductively, $E_0(I) = I$ and

$$E_{m+1}(I) = EXP(E_m(I), \varepsilon^{2^m}) \qquad m = 0, 1, 2, \ldots.$$

We shall omit the details.

Finally, assertion (vi) of the theorem may be proved by the same methods but again with substantial changes in the arithmetic. We fix $0 < p < 1$ and show that C is not (x^p)-σ-porous if the limit in (vi) vanishes. By $A_{8.3}$ this will prove the required result. To obtain a contradiction we suppose that C is (x^p)-σ-porous and therefore that a sequence $\{E_n\}$ as above may be found but this time so that the (x^p)-porosity index of E_n is at least 1 at each of its points.

In place of the estimate (1) used in the first argument we may obtain $k'(n)$ so that whenever $k \geq k'(n)$

$$\frac{(1 - 2\xi_k)^{p^n}}{[\xi_1 \xi_2 \xi_3 \cdots \xi_k]^{1-p^n}} < 3^{-n-2} \tag{8}$$

which is possible because of the limit assumed in (vi).

The induction then follows the same basic steps as used in the proofs of (i) and (iv) but using the expanded intervals

$$EXP(I_n^i, \psi)$$

where $\psi(x) = x^p$ in place of the ordinary expansion $EXP(I, 1)$. Otherwise the details, which we omit, are not much different.

Again, assertion (viii) is obtained in a similar manner. We replace statement (1) by

$$[\xi_1 \xi_2 \cdots \xi_{k-1}]^p < 3^{-n}[\xi_1 \xi_2 \cdots \xi_k] \tag{9}$$

and with other appropriate changes the proof follows the same lines.

From this theorem we obtain a number of corollaries merely by specifying the sequence ξ.

($A_{11.2}$) COROLLARY. There is a symmetric perfect set of measure zero that is not σ-porous.

PROOF. The theorem provides an example of a measure zero perfect set which is not σ-porous, by taking $\xi_n \to 1/2$ but with the series

$$\sum_{n=1}^{\infty} (1 - 2\xi_n)$$

divergent, for example $\xi_n = n/(2n+1)$. The first example of a non σ-porous measure zero perfect set was given by Zajíček [269] (although one was announced without proof by Dolženko).

($A_{11.3}$) COROLLARY. For $0 < q < 1$ there is a symmetric perfect set of measure zero that is σ-(x^q)-porous but is not σ-porous.

PROOF. In the theorem take for $p = 1/2$ the sequence

$$\xi_n = \frac{n}{2(n+2)}.$$

By ($A_{8.3}$) any σ-(x^p)-porous set is σ-(x^q)-porous.

($A_{11.4}$) COROLLARY. There is a symmetric perfect set that is porous but not σ-strongly porous.

PROOF. This follows from the theorem by taking any sequence ξ that is bounded above zero and not convergent to $1/2$. The classical Cantor ternary set provides such an example.

($A_{11.5}$) COROLLARY. Let $0 < p < 1$. There is a symmetric perfect set of positive measure that is σ-(x^p)-porous.

PROOF. For $p = 1/2$ take $\xi_n = 2^{-1}(1 - 2^{-n})$ and apply $A_{8.3}$.

($A_{11.6}$) COROLLARY. There is a symmetric perfect set of positive measure that is not σ-(x^p)-porous for any $0 < p \leq 1$.

PROOF. Take $\xi_n = 2^{-1}\left(1 - 1/(n+1)!\right)$.

($A_{11.7}$) COROLLARY. There is a symmetric perfect set that is strongly porous but not σ-(x^p)-porous for any $1 < p < +\infty$.

PROOF. Take $\xi_n = (n+2)^{-1}$ and apply the theorem.

(A$_{11.8}$) COROLLARY. For any $p > 1$ there is a symmetric perfect set that is (x^p)-porous but not σ-(x^q)-porous for any $p < q < +\infty$.

PROOF. Define an increasing sequence of numbers $\{m_n\}$ so that
$$m_1 = p, \; m_2 = p^2, \ldots, m_k = p^k.$$
Take $\xi_n = 2^{-m_n}$ and apply the theorem.

(A$_{11.9}$) COROLLARY. There is a symmetric perfect set that is (x^p)-strongly porous for all $0 < p < +\infty$.

PROOF. Define an increasing sequence of numbers $\{m_n\}$ so that $m_1 = 2$ and
$$m_{k+1} = k(m_1 + m_2 + \ldots m_k)$$
Take $\xi_n = 2^{-m_n}$ and apply the theorem.

Finally, in connection with the above results on the porosity characteristics of symmetric perfect sets, let us mention the following result which is due to Humke and Preiss [121]. They call a Borel measure μ on a compact set K a <u>Tkadlec measure</u> (after Tkadlec [241]) if there is a porous subset of K that has positive μ-measure.

(A$_{11.10}$) THEOREM. Let $C = C(\xi)$ be the symmetric perfect set given by a sequence ξ. Then every nontrivial Borel measure on C is a Tkadlec measure if and only if, for every $0 \leq s < +\infty$,
$$\sum_{n=1}^{\infty} (1 - 2\xi_n)^s = \infty.$$

REFERENCES

[1] Agronsky, S., Characterizations of certain subclasses of the Baire class 1, Doctoral Dissertation, UCSB (1974).

[2] ------------, Associated sets and continuity roads, Real Analysis Exchange 9, (1983-84), 195-205.

[3] Akemann, C.A. and Bruckner, A.M., On threading continuous functions through a compact set, Real Analysis Exchange (to appear).

[4] Baire, R., Sur les fonctions des variables réeles, Ann. Mat. Pura ed Appl. 3 (1899), 1-122.

[5] Baisnab, A.P., On a theorem of Goffman and Neugebauer, Proc. Amer. Math. Soc. 23 (1969), 573-579.

[6] --------------, On absolute Dini derivatives, Rev. Roumaine Math. Pures Appl. 15 (1970), 1593-1597.

[7] Banach, S., Sur les fonctions dérivées des fonctions mesurables, Fund. Math. 3 (1922), 128-132.

[8] ---------, Sur les ensembles de points où la dérivée est infinie, C.R.Acad. Sci. Paris 173 (1921), 457-459.

[9] Bari, N.K., A treatise on trigonometric series, Pergamon Press, New York (1964).

[10] Belna, C.L., Cluster sets of arbitrary real functions: a partial survey, Real Anal. Exchange 1 (1976), no. 1, 7-20.

[11] Belna, C.L., Cargo, G.T., Evans, M.J. and Humke, P.D., Analogues to the Denjoy-Young-Saks theorem, Proc. Amer. Math. Soc. 72 (1978), 261-267.

[12] Belna, C.L., Evans, M.J. and Humke, P.D., Symmetric and ordinary differentiation, Proc. Amer. Math. Soc. (72) 2 (1978), 261-267.

[13] ------------------------------------, Symmetric and strong differentiation, Amer. Math. Monthly 86 (2) (1979) 121-123.

[14] ------------------------------------, Symmetric monotonicity, Acta Math. Sci. Hungaricae, T. 34 (1-2) (1979) 17-22.

[15] Belowska, L., Résolution d'un problem de M. Z. Zahorski sur les limites approximatives, Fund. Math. 32 (1939), 277-286.

[16] Besicovitch, A.S., Diskussion der stetigen Funkctionen im Zusammenhang mit der Frage über ihre Differenti erbarkeit, Bull. de l'Academie des Sciences de Russie (1925), 97-122 and 527-540.

[17] -----------------, On Lipshitz numbers, Math. Zeit. 30 (1929), 514-519.

[18] -----------------, On linear sets of points of fractional dimension, Math. Ann. 101 (1929), 161-193.

[19] -----------------, Remark on relative derivatives, J. London Math. Soc. 16 (1941), 210-211.

[20] Bledsoe, W., Neighborly functions, Proc. Amer. Math. Soc. 3 (1952), 114-115.

[21] Blumberg, H., New properties of all real functions, Trans. Amer. Math. Soc. 24 (1922),

113-128.

[22] ------------, On the characterization of the set of points of λ-continuity, Annals of Math. 25 (1925), 118-122.

[23] ------------, A theorem on arbitrary functions of two variables with applications, Fund. Math. 16 (1930), 17-24.

[24] ------------, The measurable boundaries of an arbitrary function, Acta Math. 65 (1935), 263-282.

[25] Brown, J.M. and Lee, K.W., The distance set of certain Cantor sets, Real Analysis Exchange 2 (1976), 48-51.

[26] Bruckner, A.M., A theorem on monotonicity and a solution to a problem of Zahorski, Bull. Amer. Math. Soc. 71 (1965), 713-716.

[27] ------------, An affirmative answer to a problem of Zahorski, and some consequences, Michigan Math. J. 13 (1966), 15-26.

[28] ------------, On characterizing classes of functions in terms of associated sets, Canad. Math. Bull. 10 (1967), 227-231.

[29] ------------, Some remarks on extreme derivates, Canad. Math. Bull. 12 (1969), 385-388.

[30] ------------, On the differentiation of integrals in euclidean spaces, Fund. Math. 66 (1969/70), 129-135.

[31] ------------, Differentiation of integrals, Amer. Math. Monthly, 78, o. 9, part II (1971), 1-51.

[32] ------------, On everywhere convergence of sequences of convolution kernels, J. Approximation Theory 4 (1971), 218-224.

[33] ------------, <u>Differentiation of real functions</u>, Lect. Notes in Math. #659, Springer-Verlag (1978).

[34] ------------, On destruction of derivatives via changes of scale, Bull. Inst. Math. Acad. Sinica 9 (1981), 407-415.

[35] ------------, Some new simple proofs of old difficult theorems, Real Analysis Exchange 9 (1983-84), 63-78.

[36] Bruckner, A.M. and Goffman, C., The boundary behavior of real functions in the upper half plane, Rev. Roumaine Math. Pures Appl. 11 (1966), 507-518.

[37] ------------, Approximate differentiation, Real Anal. Exchange 6 (1980/81), no. 1, 9-65.

[38] Bruckner, A.M. and Haussermann, J., Strong porosity features of typical continuous functions, (to appear).

[39] ------------, Strong porosity features of typical continuous functions, Real Analysis Exchange 8 (1982-83), 21-23.

[40] Bruckner, A.M. and Johnson, K.G., Path derivatives and growth control, Proc. Amer. Math. Soc. 91 (1984), 46-48.

[41] Bruckner, A.M., Laczkovitch, M., Petruska, G. and Thomson, B.S., Porosity and approximate derivatives, (to appear).

[42] Bruckner, A.M. and Leonard, J.L., Derivatives, Amer. Math. Monthly 73 (1966), no. 4, part II, 24-56.

[43] Bruckner, A.M., O'Malley, R.J. and Thomson, B.S., Path derivatives: a unified view of certain generalized derivatives, Trans. Amer. Math. Soc., 283 (1984), 97-125.

[44] Bruckner A.M. and Thomson, B.S., Porosity estimates for the Dini derivatives, Real Analysis Exchange 9 (1983-84), 508-538.

[45] Burkill, J.C., The approximately continuous Perron integral, Math. Zeit. 34 (1931), 270-278.

[46] Burkill, J.C. and Haslam-Jones, U.S., The derivates and approximate derivates of measurable functions, Proc. London Math. Soc. 32 (1931), 346-355.

[47] Ceder, J.G., The cluster set structure of real functions, Periodica Math. Hungar.

[48] Ceder, J.G. and Pearson, T.L., A survey of Darboux Baire 1 functions, Real Analysis Exchange 9 (1983-84), 179-194.

[49] Collingwood, E.F., Cluster sets of arbitrary functions, Proc. Nat. Acad. Sci. U.S.A. 46 (1960), 1236-1242.

[50] Császár, A., Sur les nombres dérivés de Lipshitz géneralizés, Acta Math. Hungary 1 (1950), 277-302.

[51] ----------, Sur la structure des ensembles de niveau des fonctions réelles à deux variables, Acta. Sci. Math. Szewged. 15 (1954), 183-202.

[52] ----------, Sur une généralization de la notion de derivée, Acta Sc. Math. Szeged 16 (1955), 137-159.

[53] ----------, Sur la structure des ensembles de niveau des fonctions réelles à une variable, Colloq. Math. 4 (1956-57), 13-29.

[54] ----------, Sur les critères locaux de monotonite, Ann. Univ. Sci. Budapest. Eötvös Sect. Math. 5 (1962), 43-50.

[55] ----------, Sur les limites extrêmes de fonctions d'intervalle, Rev. Roumaine Math. Pures Appl. 10 (1965), 405-407.

[56] Darboux, G., Mémoire sur les fonctions discontinues, Ann. Sci. Scuola Norm. Sup. 4 (1875), 57-112.

[57] Davies, R.O., Non-monotonic implies very oscillatory, Real Analysis Exchange 6 (1980-81), 187-191.

[58] Denjoy, A., Une extension de l'intégrale de M.Lebesgue, C.R.Acad. Sci. Paris 154 (1912), 859-862.

[59] ---------, Mémoire sur les fonctions dérivées continues, Journ. Math. Pures et Appl. (7) 1 (1915), 105-240.

[60] ---------, Sur une propriété des fonctions dérivées, Enseignement Math. 18 (1916), 320-328.

[61] ---------, Mémoire sur la totalisation des nombres dérivées non sommables, Ann. Ecole Norm., 33 (1916), 127-222.

[62] ---------, Sur une propriété des séries trigonométriques, Verlag v.d. G.V. der Wis-en Natuur. Afd., 30 Oct. 1920.

[63] ─────────, Sur l'approximation de certain sommes, C.R. Acad. Sci. Paris 204 (1937), 1396-1398.

[64] ─────────, Leçons sur le calcul des coefficents d'une série trigonométrique, Part II, Métrique et topologie d'ensembles parfaits et de fonctions, Gauthier-Villars, Paris (1941).

[65] ─────────, Mémoire sur la dérivation et son calcul inverse, Gauthier-Villars (Paris) (1954).

[66] ─────────, Totalisations des dérivées premières généralisées I., C.R. Acad. Sci. Paris (241) (1955), 617-620.

[67] Dolženko, E.P., Boundary properties of arbitrary functions, Math. USSR-IZV. 1, (1967), 1-12.

[68] Evans, M.J., Qualitative aspects of differentiation, Real Analysis Exchange 9 (1983-84), 54-62.

[69] ───────────, Qualitative derivates and derivatives, Rev. Roum. Math. Pures et Appl. (to appear)

[70] ───────────, Qualitative maxima of a function, Rev. Roum. Math. Pures et Appl., (to appear).

[71] Evans, M.J. and Humke, P.D., Directional cluster sets and essential directional cluster sets of real functions defined in the upper half plane, Rev. Roum. Math. Pures et Appl. 23 (1978), 533-542.

[72] ─────────────────────────, On the approximate derivates of continuous functions, Bull. Inst. Math. Acad. Sinica 8 (1980), no. 4, 609-614.

[73] ─────────────────────────, The equality of unilateral derivates, Proc. Amer. Math. Soc. 79 (1980), no. 4, 609-613.

[74] ─────────────────────────, Parametric differentiation, Colloq. Math. 45 (1981), 125-131.

[75] Evans, M.J. and Larson L., Qualitative differentiation, Trans. Amer. Math. Soc. 280 (1983), 303-320.

[76] Ewart, J. and Lipiński, J.S., On points of continuity, quasicontinuity, and cliquishness of real functions, Real Analysis Exchange 8 (1982-83), 473-478.

[77] Foran, J. On the density maxima of a function, Colloq. Math. 37 (1977), 245-254.

[78] Foran, J. and Humke, P.D., Some set-theoretic properties of σ-porous sets, Real Anal. Exchange 6 (1980/81), no. 1, 114-119.

[79] Gál, I.S., On the fundamental theorems of the calculus, Trans. Amer. Math. Soc. 86 (1957), 309-320.

[80] Garg, K.M., An analogue of Denjoy's theorem, Ganita 12 (1961), 9-14.

[81] ─────────, On nowhere monotone functions. II. Derivates at sets of power c and at sets of positive measure, Rev. Math. Pures Appl. (Bucarest) 7 (1962), 663-671.

[82] ─────────, On nowhere monotone functions. I. Derivates at a residual set. Ann. Univ. Sci. Budapest Eötvös Sect. Math. 5 (1962), 173-177.

[83] ─────────, Applications of Denjoy analogue I (Sufficient conditions for a function to be monotone), Ann. Polonici Math. 55 (1964), 159-165.

[84] ---------, Applications of Denjoy analogue. II. Local structure of level sets and Dini derivates, Acta Math. Acad. Sci. Hungar. 14 (1963), 183-186.

[85] ---------, Applications of Denjoy analogue. III. Distribution of various typical level sets, Acta Math. Acad. Sci. Hungar. 14 (1963), 187-195.

[86] ---------, On nowhere monotone functions. III. (Functions of first and second species), Rev. Math Pures Appl. 8 (1963), 83-90.

[87] ---------, Applications of Denjoy analogue. I. (Sufficient conditions for a function to be monotone), Ann. Polon. Math. 15 (1964), 159-165.

[88] ---------, On asymmetrical derivates of non-differentiable functions, Canad. J. Math. 20 (1968), 135-143.

[89] --------, On singular functions, Rev. Roumaine Math. Pures Appl. 14 (1969), 1441-1452.

[90] ---------, On a residual set of continuous functions, Czechoslovak Math. J. 20 (95) (1970), 537-543.

[91] ---------, Characterizations of absolutely continuous and singular functions, Proc. of the Conference on the Constructive Theory of Functions (Approximation Theory) (Budapest, 1969), pp. 183-188. Akademiai Kiado, Budapest, 1972.

[92] ---------, Monotonicity, continuity and levels of Darboux functions, Colloq. Math. 28 (1973), 91-103, 162.

[93] ---------, On bilateral derivates and the derivative, Trans. Amer. Math. Soc. 210 (1975), 295-329.

[94] ---------, On a new definition of derivative, Bull. Amer. Math. Soc. 82 (1976), no. 5, 768-770.

[95] ---------, A new notion of derivative, Real Anal. Exchange, 7 (1981/82), 65-84.

[96] --------, On derivatives of variation functions, The Eoin L.Whitney Memorial Collection, University of Alberta (1967).

[97] --------, Derivatives of variation functions, of mutually singular functions, and relatively absolutely continuous functions, Real Analysis Exchange 10 (1984-85),pp.54-56.

[98] Gleyzal, A., Interval-functions, Duke Math. J. 8 (1941), 223-230.

[99] Goffman, C., On Lebesgue's density theorem, Proc. Amer. Math. Soc. 1 (1950), 384-387.

[100] -----------, A generalization of the Riemann integral, Proc. Amer. Math. Soc. 3 (1952), 543-547.

[101] -----------, On the approximate limits of real functions, Acta Sci. Math. 23 (1962), 76-87.

[102] Goffman, C. and Neugebauer, C.J., On approximate derivatives, Proc. Amer. Math. Soc. 11 (1960), 962-966.

[103] Goffman, C., Neugebauer, C.J., and Nishiura, T., Density topology and approximate continuity,

[104] Goffman, C. and Waterman, D., Approximately continuous transformations, Proc. Amer. Math. Soc. 12 (1961), 116-121.

[105] Goldowsky, G., Note sur les dérivées éxactes, Rec. Math. Soc. Math. Moscou 35 (1928), 35-36.

[106] Good, I.J., The approximate local monotony of measurable functions, Proc. Camb. Phil. Soc. 36 (1940), 9-13.

[107] Hájek, O., Note sur la mesurabilité B de la dérivée supérieure, Fund. Math. 44 (1957), 238-140.

[108] Hanson, E.H., A new proof of the theorem of Denjoy, Young and Saks, Bull. Amer. Math. Soc. (2) 40 (1934), 691-694.

[109] Haussermann, J., Porosity characterizations of intersection sets with the typical continuous function, Real Analysis Exchange 9 (1983-84), 386-389.

[110] --------------, Thesis, U.C.S.B. (in preparation).

[111] Heindl,G. and Köhler,G., Ein Montoniekriterium, Bayer. Akad. Wiss. Math.-Natur. Kl. S.-B. 1968, Abt. II, 107-112 (1969).

[112] Henstock,R., The equivalence of generalized forms of the Ward, Variational, Denjoy-Stieltjes, and Perron-Stieltjes integrals, Proc. Lond. Math. Soc. (3) 10 (1960), 281-303.

[113] -----------, Theory of integration, London, Butterworths (1963).

[114] -----------, Linear analysis, Butterworths, London (1968).

[115] -----------, Generalized integrals of vector-valued functions, Proc. London Math. Soc. (3) 19 (1969), 509-536.

[116] -----------, The variation on the real line, Proc. Roy. Irish Acad. Sect. A 79 (1979), 1-10.

[117] Hilbert, D., Grundzüge einer allgemeinen Theorie der linearen Integralgleichungen, vierte Mitteilung, (1906) Gött. Nachr., p. 168.

[118] Humke, P.D., Nowhere monotone functions and a question of K.M. Garg, Fund. Math.

[119] ----------, Some problems in need of solution, Real Anal. Exch., 7 (1981/82), 31-41.

[120] ----------, Some remarks on σ-porous sets and unilateral derivates, Real Analysis Exchange 8 (1982-83), 33-34.

[121] Humke, P.D., and Preiss, D., Measures for which σ-porous sets are null, J. Lond. Math. Soc. (to appear).

[122] Humke, P.D. and Thomson, B.S., A porosity characterization of symmetric perfect sets,(to appear).

[123] Humke, P.D. and Vessey, T., Another note on σ-porous sets, Real Analysis Exchange 8 (1982-83), 261-271.

[124] Hunter, U., Essential cluster sets, Trans. Amer. Math. Soc. 119 (1965), 380-388.

[125] ----------, An abstract formulation of some theorems on cluster sets, Proc. Amer. Math. Soc. 16 (1965), 909-912.

[126] Jarník, V., Sur les fonctions de deux variables réelles, Fund. Math. 27 (1936), 147-150.

[127] Jędrzejewski, J.M., Approximately smooth functions. (Polish), Zeszyty Nauk. Univ. Lodz. Nauki Mat. Przyrod. Ser. II Zeszyt 52 Mat. (1973), 7-14.

[128] -----------------, The generalized limit and generalized continuity. (Polish), Zeszyty Nauk Univ. Lodz. Nauki. Mat. Przyrod. Ser. II Zeszyt 52 Mat. (1973), 19-38.

[129] ----------------, On limit numbers of real functions, Fund. Math. 83 (1973/74), no. 3, 269-281.

[130] Jędrzejewski, J.M. and Wilczyński, W., On the family of sets of limit numbers, Bull. Acad. Polon. Sci. Ser. Sci. Math. Astronom. Phys. 18 (1970), 453-460.

[131] ---------------------------------, The family of sets of limit numbers, Zeszyty Nauk. Univ. Lodz. Nauki Mat. Przyrod. Ser. II Zeszyt 52 Mat. (1973), 39-43.

[132] Jeffery, R.L., The theory of functions of a real variable, Math. Expositions #6, U. of T. Press, 1962.

[133] Jurek, B., Sur les nombres dérivés de fonctions discontinues, Ceska Spol. Nauk Trida Math. Prirodovedecka Vestnik, 1 (1937), 1-22, Prirod.

[134] Kempisty, S., Sur les fonctions quasicontinues, Fund. Math. 19 (1932), 184-197.

[135] Khintchine, A., Sur la dérivation asymptotique, C.R Acad. Sci. Paris, 164 (1917), 142-144.

[136] -------------, An investigation of the structure of measurable functions, (in Russian) Mat. Sbornik 31 (1924), 265-285.

[137] -------------, Recherches sur la structure des fonctions mesurables, Fund. Math. 9 (1927), 212-279.

[138] Koszela, B., Świątkowski, T. and Wilczyński, W., Classes of continuous real functions, Real Analysis Exchange 4 (1978-79), 139-157.

[139] Kulbacka, M., Sur l'ensemble des points de l'assymetrie approximative, Acta. Sci. Math. 21 (1960), 90-95.

[140] Kuratowski, K., Topology, Academic Press (1966), London.

[141] Laczkovich, M., On the Baire class of the upper derivatives, Anal. Math. 1 (1975), no. 2, 115-120.

[142] -------------, On the Baire class of selective derivatives, Acta Math. Acad. Sci. Hungar. 29 (1977), no. 1 -2, 99-105.

[143] -------------, Baire 1 functions, Real Analysis Exchange 9 (1983-84), 15-28.

[144] Laczkovich, M. and Petruska, G., Baire 1 functions, approximately continuous functions, and derivatives, Actá Math. Acad. Sci. Hungar. 25 (1974), 189-212.

[145] ----------------------------, Remarks on a problem of A.M. Bruckner, Real Analysis Exchange 6 (1980/81), 120-126

[146] ----------------------------, Remarks on a problem of A.M. Bruckner, Acta. Math. Acad. Sci. Hungar. 38 (1981), 205-214.

[147] Laczkovich, M. and Preiss, D., α-variation and transformation into C^n functions, Real Analysis Exchange 10 (1984/1985), 21- 24.

[148] Levi, B., Richerche sulle funzioni derivate, Rend. dei Lincei (5),vol.XV (1906), p.437.

[149] Lipiński, J.S., Sur la classe M_3', Casopis Pest. Mat. 93 (1968), 222-226, 227.

[150] Lipiński, J.S., Sur la discontinuité approximative et la dérivée approximative, Colloq. Math. 10 (1963), 103-109.

[151] Liu, Fon-Che, Approximation-extension type proprerty of continuous functions of bounded variation, Journ. Math. Mech. 19 (1969), 207-218.

[152] Leonard, J.L., Some conditions implying the monotonicity of a real function, Rev. Roumaine Math. Pures Appl. 17 (197 2), 757-780.

[153] Luczak, On some problems concerning a generalized derivative of a real function, Zeszyty Nauk Politech. Lodz. Mat. no. 12 (1979), 5-13.

[154] Marcus, S., Approximately qualitative limit (in Roumanian), Comun. Acad. RPR, 3 (1953), 9-12.

[155] ----------, Approximately qualitative continuity (in Roumanian), Comun. Acad. RPR, 3 (1953), 117-120.

[156] ----------, An analysis of real-valued functions based on the notion of category in Baire's sense (in Roumanian), Studii Cerc. Mat. 7 (1956), 251-272.

[157] ----------, Sur les fonctions quasicontinues, Colloq. Math., 8 (1961), 47-53.

[158] Mastalerz-Wawrzyńczak, M., On a certain condition of the monotonicity of functions, Fund. Math. 97 (1977), no. 3, 18 7-198.

[159] Matysiak, A., Sur les limites approximatives, Fund. Math. 48 (1960), 363-366.

[160] Mawhin, J., Introduction à l'analyse, Cabay, Louvain-la-Neuve, 1979.

[161] McGill, P., Properties of the variation, Proc. Roy. Irish Acad. Sect. A 75 (1975), 73-77.

[162] Mišík, L., Über die Klasse M_3, Casopis Pest. Mat. 91 (1966), 389-393.

[163] --------, Notes on an approximate derivative, Mat. Casopis Sloven. Akad. Vied 22 (1972), 108-114.

[164] --------, Über einen Satz von Khintchine. II. Mat. Casopis Sloven. Akad. Vied 24 (1974), 145-154.

[165] --------, Über approximative derivierte Zahlen, Czechoslovak Math. J. 25 (100) (1975), 154-159.

[166] --------, Über approximative derivierte Zahlen monotoner Funktionen, Czechoslovak Math. J. 26 (101) (1976), no. 4, 579-583.

[167] Monroe, M.E., Measure and integration, 2nd Ed., Addison-Wesley (1971).

[168] Morse, A.P., Dini derivatives of continuous functions, Proc. Amer. Math. Soc. 5 (1954), 126-130.

[169] Natanson, I., Theory of functions of a real variable, 2 vols., Ungar, New York (1961).

[170] Neubrunnova, A., On quasicontinuous and cliquish functions, Casopis Pest. mat. 99 (1974), 109-114.

[171] Neugebauer, C.J., A theorem on derivates, Acta. Sci. Math. (Szeged) 23 (1962), 79-81.

[172] ----------------, Darboux functions of Baire class one and derivatives, Proc. Amer. Math. Soc. 13 (1962), 838-843

[173] Nishiura, T., Remarks on Agronsky's theorem on strong containment and continuity roads, Real Analysis Exchange 9 (1983-84), 400-414.

[174] O'Malley, R,J., Strict essential maxima, Proc. Amer. Math. Soc. 33 (1972), 501-504.

[175] -------------, A density property and applications Trans. Amer. Math. Soc. 199 (1974), 75-87.

[176] ------------, M_2 functions, Indiana Univ. Math. J. 24 (1974/75), 585-591.

[177] ------------, Baire* 1, Darboux functions, Proc. Amer. Math. Soc. 60 (1976), 187-192.

[178] ------------, The set where an approximate derivative is a derivative, Proc. Amer. Math. Soc. 54 (1976), 122-124.

[179] ------------, Note about preponderantly continuou functions, Rev. Roumaine Math. Pures Appl. 21 (1976), 335-336.

[180] ------------, Selective derivates, Acta. Math. Acad. Sci. Hungar. 29 (1977), 77-97.

[181] ------------, Approximately differentiable functions: the r topology, Pacific J. Math. 72 (1977), 207-222.

[182] ------------, Aproximate maxima, Fund. Math. 94 (1977), 75-81.

[183] ------------, Decomposition of approximate derivatives, Proc. Amer. Math. Soc. 69 (1978), 234-247.

[184] ------------, Approximately continuous functions which are continuous almost everywhere, Acta. Math. Acad. Sci. Hungar. 33 (1979), 395-402.

[185] ------------, Insertion of Baire* 1, Darboux functions, Rev. Roumanine Math. Pures Appl. 24 (1979), 1445-1448.

[186] ------------, Selective derivatives and the Denjoy-Clarkson properties, Acta Math. Acad. Sci. Hungar. 36 (1980), 195-199.

[187] ------------, Balanced selections, Real Analysis Exchange 8 (1982-83), 504-508.

[188] ------------, The multiple intersection property for path derivatives, Real Analysis Exchange 9 (1983-84), 32-34.

[189] O'Malley, R.J. and Weil, C.E., The oscillatory behaviour of certain derivatives, Trans. Amer. Math. Soc. 234 (1977), 467-481.

[190] ---------------------------, Selective, bi-selective, and composite differentiation, Acta Math. Acad. Sci. Hungar. (to appear).

[191] Ornstein, D., A characterization of monotone functions, Illinois J. Math. 15 (1971), 73-76.

[192] Ostaszewski, K., Continuity in the density topology, Real Analysis Exchange 7 (1981-82), 259-270.

[193] ----------------, Density topology and the Luzin (N) condition, Real Analysis Exchange 9 (1983-84), 390-393.

[194] Oxtoby, J., Measure and category, Springer-Verlag (1971).

[195] Preiss, D., Approximate derivatives and Baire classes, Czech. Math. J. (21) 96 (1971), 373-382.

[196] ----------, Limits of approximately continuous functions, Czech. Math. J. (21) 96 (1971), 371-372.

[197] ----------, Level sets of derivatives, Trans. Amer. Math. Soc. 272 (1982), 161-184.

[198] Preiss, D. and Zajíček, L., On the symmetry of approximate Dini derivates of arbitrary functions, Comment. Math. Univ. Carolinae 6 (1980), 77-93.

[199] Pu, H.W., A monotonicity criterion for arbitrary functions, Colloq. Math. 31 (1974), 289-292.

[200] -------, Associated sets of Baire* 1 functions, Real Analysis Exchange 8 (1982-83), 479-485.

[201] Pu, H.W. and Pu, H.H., On the approximate maxima of a function, Rev. Roum. Math. Pures et Appl. 24 (1979), 281-284.

[202] Pu, H.H.,Chen, J.D., and Pu, H.W., A theorem on approximate derivates, Bull. Inst. Math. Acad. Sinica 2 (1974), 87-91.

[203] Ravetz, J., The Denjoy theorem and sets of fractional dimension, J. London Math. Soc. 29 (1954), 88-96.

[204] Redheffer, R., Increasing functions, Aequat. Math. 22 (1981), 119-133.

[205] Rinne, D., Characterizing cluster sets of real functions, Real Analysis Exchange 5 (1979-80), 164-179.

[206] Rogers, C.A., Hausdorff measures, Cambridge University Press (1970).

[207] Rosenthal, A., Über die Singularitäten der reellen ebenen Kurven, Habilitationschrift, München (1912).

[208] Saks, S., Sur les nombres dérivés des fonctions, Fund. Math. 5 (1924), 98-104.

[209] Saks, S., Theory of the integral, Monografie Matematyczne 7, Warsawa-Lwow, (1937).

[210] Scheefer, L., Zur Theorie der stetigen Funktionen einer reellen Veränderlichen, Acta. Math. 5 (1884/85), 279-296.

[211] Shukla, P.D., On the differentiability of monotone functions, Bull. Calcutta Math.Soc. 37 (1945), 9-14.

[212] Sierpiński, W., Sur l'ensemble des points angulaires d'une courbe y=f(x), Bull. Acad. Sci. Cracovie (1912), 850-855.

[213] -------------, Un lemma métrique, Fund. Math. 4 (1923), 201-203.

[214] Sindalovskiĭ,G.H., Continuity and differentiability with respect to congruent sets, Soviet Math. Dokl. 1 (1961), 1217-1218.

[215] -----------------, On a generalization of derived numbers, Izv. Akad. Nauk SSSR Ser. Mat. 24 (1960), 707-720.

[216] -----------------, Continuity and differentiability with respect to congruent sets, Izv. Akad. Nauk SSSR Ser. Mat. 26 (1962), 125-142.

[217] -----------------, Congruent and asymptotic differentiability, Dokl. Akad. Nauk SSSR 150 (1963), 995-997.

[218] -----------------, Differentiability with respect to congruent sets, Izv. Akad. Nauk SSSR Ser. Mat. 29 (1965), 11-40.

[219] -----------------, On the equivalence between ordinary derivatives and the derivatives with respect to congruent sets of a certain class, Izv. Akad. Nauk SSSR Ser. Mat. 29 (1965) 987-996.

[220] -----------------, The derived numbers of continuous functions, Uspehi Mat. Nauk 21 (1966), 274-277.

[221] ----------------, The derived numbers of continuous functions, Math. USSR Izv. 2 (1968), 943-978.

[222] Snyder, L.E., Approximate Stolz angle limits, Proc. Amer. Math. Soc. 17 (1966), 416-422.

[223] Starcev, V.A., Symmetyric continuity and symmetric differentiability with respect to sets, Soviet Math. Dokl. 10 (1969), 517-519.

[224] Sunouchi, G. and Utagawa, M., The generalized Perron integrals, Tohoku Math. J. (2) 1 (1949), 95-99.

[225] Świątkowski, T., On the conditions of monotonicity of functions, Fund. Math. 59 (1966), 189-201.

[226] --------------, On some generalizations of the notion of asymmetry of functions, Coll. Math. 17 (1967), 77-91.

[227] --------------, On a certain generalization of the notion of derivative (Polish), Zeszyty Naukowe Politech. Lodzkiej 149, Mat. 2.1 (1972), 89-103.

[228] Tall, F.D., The density topology, Pacific J. Math. 62 (1976), 275-284.

[229] Tevy, I., Topologies engendrées par des idéaux, Rev. Roum. Math. Pures et Appl., (to appear).

[230] Tevy, I. and Bruteanu, C., On some continuity notions, Rev. Roum. Math., Pures et Appl., 18 (1973), 121-135.

[231] Theilman, H.P., Types of functions, Amer. Math. Monthly 60 (1953), 156-161.

[232] Thomson, B.S., On the total variation of a function, Canad. Math. Bull.

[233] ------------, Outer measures and total variation, Canad. Math. Bull.

[234] ------------, Monotonicity theorems, Proc. Amer. Math. Soc. 83 (1981), 547-552.

[235] ------------, Monotonicity theorems, Real Analysis Exchange 6 (1981), 209-234.

[236] ------------, Some theorems for extreme derivates, J. London Math. Soc. (to appear)

[237] ------------, Some properties of generalized derivatives, Real Analysis Exchange 8 (1982-83), 58-59.

[238] ------------, Derivation bases on the real line, I and II, Real Analysis Exchange, 8 (1982-83), pp.67-208 and pp.278-442.

[239] -----------, On the level set structure of a continuous function (to appear).

[240] Tkadlec, J., Construction of some non σ-porous sets on the real line, Real Analysis Exchange 9 (1983-84), 473-482.

[241] -----------, Measures on nonporous sets, (to appear),

[242] Tolstov, G., Sur quelques propriétés des fonctions approximativement continues, Rec. Mat. (Mat. Sbornik) N.S., 5 (1939), 637-645.

[243] Tonelli, L., Sulle derivate esatte, Mem. Istit. Bologna (8) 8 (1930/31), 13-15.

[244] Vessey, T., On porosity and exceptional sets, Real Analysis Exchange 9 (1983-84), 336-340.

[245] Ward, A.J., On the points where $AD^+ < AD^-$, J. London Math. Soc. 8 (1933), 293-299.

[246] ---------, On the differential structure of real functions, Proc. London Math. Soc. 39 (1935), 339-362.

[247] Ważewski, T., Sur une condition nécessaire et suffisante pour qu'une fonction continue soit monotone, Ann. Soc. Polon. Math. 24 (1951), 111-119.

[248] Weil, C.E., On properties of derivatives, Trans. Amer. Math. Soc. 114 (1965), 363-376.

[249] ---------, On approximate and Peano derivatives, Proc. Amer. Math. Soc. 20 (1969), 487-490.

[250] ---------, A property for certain derivatives, Indiana Univ. Math. J. 23 (1973/74), 527-536.

[251] ---------, Monotonicity, convexity and symmetric derivates, Trans. Amer. Math. Soc. 221 (1976), 225-237.

[252] ---------, On nowhere monotone functions, Proc. Amer. Math. Soc. 56 (1976), 388-389.

[253] Willard, S., General Topology, Addison-Wesley (1970).

[254] Wilczyński, W., On the family of sets of approximate limit numbers, Fund. Math. 75 (1972), 169-174.

[255] -------------, A generalization of the density topology, Real Analysis Exchange 8 (1982), 16-20.

[256] Yanagihara, N., Angular cluster sets and horicyclic cluster sets, Proc. Japa Acad. 45 (1969),427-428.

[257] Young, G.C., On the derivates of a function, Proc. London Math. Soc. (2) 15 (1916), 360-384.

[258] Young, W.H., Oscillating successions of continuous functions, Proc. London Math. Soc. (2) 6 (1908), 298-320.

[259] ----------, On the distinction of right and left a points of discontinuity, Quart. J. Pure Appl. Math. 39 (1908), 67-83.

[260] ----------, On some applications of semicontinuous functions, Atti del IV Congresso Intern. dei Mathematici II (Roma, 1909), 49-60.

[261] ----------, On the discontinuities of a function of one or more real variables, Proc. London Math. Soc. (2) 8 (1909), 117-124.

[262] ----------, On integration with respect to a function of bounded variation, Proc. London Math. Soc. 13 (1914), 109-150.

[263] ----------, La symétrie de structure des fonctions des variables réelles, Bull. Sci. Math. (2) 52 (1928), 265-280.

[264] Young, W.H. and Young, G.C., On the reduction of sets of intervals, Proc. London Math. Soc.,(2) 14 (1915), 111-130

[265] Zahorski, Z., Sur la première derivée, Trans. Amer Math. Soc. 69 (1950), 1-54.

[266] Zajíček, L., On Baire classes, Casopis Pest. Mat. 9 (1970), 240-241.

[267] ----------, On the intersection of the sets of the right and left internal approximate derivatives, Czech. Math. J. 23 (98) (1973), 521-526.

[268] ----------, On cluster sets of arbitrary functions Fund. Math. 83 (1973/74), 197-217.

[269] ----------, Sets of σ-porosity and sets of σ-porosity (q), Casopis Pest. Mat. 101 (1976),350-359.

[270] ----------, On the symmetry of Dini derivates of arbitrary functions, Comment. Math. Univ. Carolin. 22 (1981), 195-209.

[271] ----------, On Dini derivatives of continuous and monotone functions, Real Analysis Exchange 7 (1981/82), 233-238.

[272] ----------, On approximate Dini derivatives and one-sided approximate derivatives of arbitrary functions, Comment. math. Univ. Carolin. 22 (1981), 549-560.

[273] ----------, A note on preponderant maxima, Colloq. Math.

LIST OF SPECIAL SYMBOLS

| | (exterior Lebesgue measure)
| |^i (interior Lebesgue measure)

ACG, 129
ACG*, 105, 122

C(ψ), 125, 157-160, 203-205

d(x, I), 185
$\overline{D} f(x)$, $\underline{D} f(x)$ (bilateral derivates), 15
$\overline{D}_{ap} f(x)$, $\underline{D}_{ap} f(x)$ (approximate derivates), 15
$\overline{D}^+ f(x)$, $\underline{D}^+ f(x)$ (Dini derivatives), 15
$\overline{D}^\# f(x)$, $\underline{D}^\# f(x)$ (sharp derivates), 15
$\overline{D}_N f(x)$, $\underline{D}_N f(x)$, 15
Δf, 39

EXP(I, ψ), 187

$[J_1]$, 10, 48-50
$[J_2]$, 29
$[J_3]$, 81
$[J_4]$, 82

$\lambda(E, x, y)$, 184
λf (Lebesgue-Stieltjes measure), 96
$\Lambda(f, x)$, 45
$\Lambda^+(f, x)$, $\Lambda^-(f, x)$, 45
Λess(f, x), 46
$\Lambda(f, x)$, 45

M_2-property, 72
M_3-property, 55, 72, 195
M_4-property, 195

$PI^+(E; x)$, $PI^-(E; x)$, 185
$PI_\psi(E; x)$, 197

N-continuous, 72

ψ-porosity, 197
ψ-porosity index, 197

R (real numbers)
$\overline{\mathbf{R}}$ (extended real numbers)

S (local system), 3
S_0, 3-4,
S_0^+, 5

S_0^-, 5
$S_1 \vee S_2$, 5
$S_1 \wedge S_2$, 5
$S_1 \ll S_2$, 5
S_{ap}, 22
S_N, 21
S_u, 19
S_{wap}, 22
S_∞, 3-4
S_τ, 18
S^* (dual system), 7
S-μ_ψ, 41
(S)-$\Lambda(f, x)$ (cluster set), 3
(S)-continuous, 70, 89
(S)-cover, 35-38, 40, 41
(S)-derived set, 50
(S)-$\overline{D} f(x)$ (derivate), 14
(S)-lim $_{y \to x}$ (point function limit), 3
(S)-lim $_{I \to x}$ (interval function limit), 3
(S)-limsup $_{y \to x}$, 3
(S)-liminf $_{y \to x}$, 3
(S)-negligible, 51
σ-(ψ)-porosity, 200
σ-porous, 55, 65-66, 184, 189-190
σ-(S)-negligible, 51, 53
σ-strongly-porous, 185

u-derivative, 19

$V(\psi, S; X)$, 39
$V_K(\psi, S; X)$, 41
$Var(\psi, C)$, 38
$Var_K(\psi, C)$, 41
VBG, 95, 122
[VBG], 95
VBG*, 93-95, 97-98, 103-104, 106-107, 113, 115, 122, 130, 134

$(x\alpha)$-porosity, 197

$[Z_\lambda]$, 127-129, 150

INDEX

ACG, 129
ACG*, 105, 122
absolutely continuous, 105-106
ambiguity theorem, 44, 62
approximate bilateral derivate, 134
approximate continuity, 73, 148
approximate derivative, 24, 155, 160, 162-165
approximate Dini derivatives, 15, 24, 129, 135-136, 139, 145, 147-150, 153, 155, 170-175
approximate system, 22
approximately constant, 118-119
approximately increasing, 117-119
approximately oscillatory, 118-119
asymmetry theorem, 44, 56, 60-62

Baire property, 79
Baire 1 function, 16, 72, 74-77, 80, 90, 137, 196
Baire 2 function, 52
Beppo Levi theorem, 143
bilateral, 37
bilateral approximate derivate, 24
bilateral derivates, 14
bilaterally increasing, 115-116
bilaterally strongly porous, 184

$C(\psi)$, 125, 157-160, 203-205
Cauchy Schwartz inequality, 111
Cantor set, 64-65, 206, 210
closure operator, 19-21, 47, 56
cluster set, 3, 6, 24, 44-69, 183
congruent system, 25, 162-163
cover, 35

Darboux function, 77
Denjoy-Khintchine theorem, 160
Denjoy-Young-Saks theorem, 139, 166-182
density system, 22-24, 27-28, 31, 33-34, 52, 54
density topology, 19, 23
derived set, 50
Dini derivatives, 15, 122, 135, 138, 166, 115
dual properties, 10
dual system, 7

$EXP(I, \psi)$, 187
essential asymmetry, 64-67

essential cluster set, 44, 46, 64, 76-69
Evans-Humke porosity lemma, 191
Evans-Humke theorem, 158-160
extreme bilateral derivates, 14
extreme derivates, 14

feeble continuity, 53, 70
filtering, 10
finite variation, 39, 86, 90-96, 110
full cover, 35
full measure system, 25

Goldowsky-Tonelli theorem, 137
guage, 32

Hausdorff measure, 202-203

increasing sets property, 41-43
intersection condition, 31-37, 74, 88, 91, 101-121, 128-129, 143-145, 151, 158, 162

$[J_1]$, 10, 48-50
$[J_2]$, 29
$[J_3]$, 81
$[J_4]$, 82
joint intersection condition, 59-61, 63-64
Jordan decomposition, 92

$\lambda(E, a, b)$, 184
lattice, 5
Lebesgue decomposition, 93
Lebesgue Stieltjes measure, 96, 100, 105
local system, 3
locally increasing, 115

M_2-property, 72
M_3-property, 55, 72, 195
M_4-property, 195
measure continuous, 78
metric outer measure, 40
monotonicity property, 131
mutually singular functions, 110-114

N-continuous, 72
nearly continuous, 78
negligent cluster set, 62
negligent continuity, 78-80
negligent Dini derivatives, 22, 129, 136,

145-146, 150, 179-182
negligent limits, 21-22
nonporous, 184

outer measure, 40

ψ-porosity, 197
ψ-porosity index, 197
partial order, 5
path generated, 30
path limit, 27-31
path system, 25, 30
porosity derivate, 194
porosity derivates, 26, 161
porosity function, 197
porosity index, 185
porosity lemma, 191-194, 203-205
porosity system, 26
porosity, 26, 55, 65, 72, 119, 123-127, 139, 148, 151, 153, 155-157, 158-160, 161, 163, 172-177, 179-182, 183-211
porous, 184
positive measure system, 24, 46, 67
preponderant continuity, 73

qualitative continuity, 78
qualitative Dini derivatives, 22, 179
quasi limit, 26, 82

relative monotonicity, 115, 120-130

S_0, 3-4,
S_0^+, 5
S_0^-, 5
$S_1 \vee S_2$, 5
$S_1 \wedge S_2$, 5
S_{ap}, 22
S_N, 21
S_{wap}, 22
S_∞, 3-4
S_T, 18
S^*, 7
(S)-cluster set, 3
(S)-continuous, 70, 89
(S)-cover, 35-38, 40, 41
(S)-derived set, 50
(S)-limit, 3
(S)-negligible, 51
σ-(ψ)-porosity, 200
σ-porous, 55, 65-66, 184, 189-190
σ-(S)-negligible, 51, 53

σ-strongly-porous, 185
Saks-Sierpiński theorem, 101
selective system, 26
semicontinuous, 71, 74
sequential derivative, 164-165
sequential system, 25
sharp derivates, 15, 138, 151, 154-155, 178
simply approached, 44-45
singular function, 108-114
Stieltjes measure, 39, 85
strict limit, 12-13
strongly porous, 184
symmetric perfect set, 205-211

Tkadlec measure, 211
theorem,
    of Beppo Levi, 138, 143, 147
    of Burkill, 136
    of de la Vallée Poussin, 106
    of Denjoy-Khintchine, 160
    of Denjoy, Young and Saks, 166-182
    of Evans-Humke, 158-161
    of Goldowski and Tonelli, 137
    of Humke and Preiss, 211
    of O'Malley, 136
    of Saks and Sierpiński, 101
    of Titchmarch and Aumann, 135
    of Vitali, 96-98, 167, 170
    of Ważewski, 135
    of G.C. Young, 138, 143, 146-147
    of W.H. Young, 138, 151-155, 160, 178

u-derivative, 19
ultrafilters, 11

VBG, 95, 122
[VBG], 95
VBG*, 93-95, 97-98, 103-104, 106-107, 113, 115, 122, 130, 134
variation, 38, 85-114
variationally equivalent, 39, 86-87, 92, 104
Vitali cover, 36
Vitali theorem, 96-98, 167, 170

W.H. Young theorem, 138, 151-155, 159, 178
weak approximate system, 22
weakly approximately continuous, 54
weakly approximately increasing, 117-119
weakly N-approached, 47
weakly N-continuous, 54, 73

(x^α)-porosity, 197

$[Z\lambda]$, 127-129, 150
Zahorski derivate, 194
zero variation, 39, 86

Vol. 1008: Algebraic Geometry. Proceedings, 1981. Edited by J. Dolgachev. V, 138 pages. 1983.

Vol. 1009: T. A. Chapman, Controlled Simple Homotopy Theory and Applications. III, 94 pages. 1983.

Vol. 1010: J.-E. Dies, Chaînes de Markov sur les permutations. IX, 226 pages. 1983.

Vol. 1011: J. M. Sigal. Scattering Theory for Many-Body Quantum Mechanical Systems. IV, 132 pages. 1983.

Vol. 1012: S. Kantorovitz, Spectral Theory of Banach Space Operators. V, 179 pages. 1983.

Vol. 1013: Complex Analysis – Fifth Romanian-Finnish Seminar. Part 1. Proceedings, 1981. Edited by C. Andreian Cazacu, N. Boboc, M. Jurchescu and I. Suciu. XX, 393 pages. 1983.

Vol. 1014: Complex Analysis – Fifth Romanian-Finnish Seminar. Part 2. Proceedings, 1981. Edited by C. Andreian Cazacu, N. Boboc, M. Jurchescu and I. Suciu. XX, 334 pages. 1983.

Vol. 1015: Equations différentielles et systèmes de Pfaff dans le champ complexe – II. Seminar. Edited by R. Gérard et J. P. Ramis. V, 411 pages. 1983.

Vol. 1016: Algebraic Geometry. Proceedings, 1982. Edited by M. Raynaud and T. Shioda. VIII, 528 pages. 1983.

Vol. 1017: Equadiff 82. Proceedings, 1982. Edited by H. W. Knobloch and K. Schmitt. XXIII, 666 pages. 1983.

Vol. 1018: Graph Theory, Łagów 1981. Proceedings. Edited by M. Borowiecki, J. W. Kennedy and M. M. Sysło. X, 289 pages. 1983.

Vol. 1019: Cabal Seminar 79–81. Proceedings, 1979–81. Edited by A. S. Kechris, D. A. Martin and Y. N. Moschovakis. V, 284 pages. 1983.

Vol. 1020: Non Commutative Harmonic Analysis and Lie Groups. Proceedings, 1982. Edited by J. Carmona and M. Vergne. V, 187 pages. 1983.

Vol. 1021: Probability Theory and Mathematical Statistics. Proceedings, 1982. Edited by K. Itô and J.V. Prokhorov. VIII, 747 pages. 1983.

Vol. 1022: G. Gentili, S. Salamon and J.-P. Vigué. Geometry Seminar "Luigi Bianchi", 1982. Edited by E. Vesentini. VI, 177 pages. 1983.

Vol. 1023: S. McAdam, Asymptotic Prime Divisors. IX, 118 pages. 1983.

Vol. 1024: Lie Group Representations I. Proceedings, 1982–1983. Edited by R. Herb, R. Lipsman and J. Rosenberg. IX, 369 pages. 1983.

Vol. 1025: D. Tanré, Homotopie Rationnelle: Modèles de Chen, Quillen, Sullivan. X, 211 pages. 1983.

Vol. 1026: W. Plesken, Group Rings of Finite Groups Over p-adic Integers. V, 151 pages. 1983.

Vol. 1027: M. Hasumi, Hardy Classes on Infinitely Connected Riemann Surfaces. XII, 280 pages. 1983.

Vol. 1028: Séminaire d'Analyse P. Lelong – P. Dolbeault – H. Skoda. Années 1981/1983. Edité par P. Lelong, P. Dolbeault et H. Skoda. VIII, 328 pages. 1983.

Vol. 1029: Séminaire d'Algèbre Paul Dubreil et Marie-Paule Malliavin. Proceedings, 1982. Edité par M.-P. Malliavin. V, 339 pages. 1983.

Vol. 1030: U. Christian, Selberg's Zeta-, L-, and Eisensteinseries. XII, 196 pages. 1983.

Vol. 1031: Dynamics and Processes. Proceedings, 1981. Edited by Ph. Blanchard and L. Streit. IX, 213 pages. 1983.

Vol. 1032: Ordinary Differential Equations and Operators. Proceedings, 1982. Edited by W. N. Everitt and R. T. Lewis. XV, 521 pages. 1983.

Vol. 1033: Measure Theory and its Applications. Proceedings, 1982. Edited by J. M. Belley, J. Dubois and P. Morales. XV, 317 pages. 1983.

Vol. 1034: J. Musielak, Orlicz Spaces and Modular Spaces. V, 222 pages. 1983.

Vol. 1035: The Mathematics and Physics of Disordered Media. Proceedings, 1983. Edited by B. D. Hughes and B. W. Ninham. VII, 432 pages. 1983.

Vol. 1036: Combinatorial Mathematics X. Proceedings, 1982. Edited by L. R. A. Casse. XI, 419 pages. 1983.

Vol. 1037: Non-linear Partial Differential Operators and Quantization Procedures. Proceedings, 1981. Edited by S. I. Andersson and H.-D. Doebner. VII, 334 pages. 1983.

Vol. 1038: F. Borceux, G. Van den Bossche, Algebra in a Localic Topos with Applications to Ring Theory. IX, 240 pages. 1983.

Vol. 1039: Analytic Functions, Błażejewko 1982. Proceedings. Edited by J. Ławrynowicz. X, 494 pages. 1983

Vol. 1040: A. Good, Local Analysis of Selberg's Trace Formula. III, 128 pages. 1983.

Vol. 1041: Lie Group Representations II. Proceedings 1982–1983. Edited by R. Herb, S. Kudla, R. Lipsman and J. Rosenberg. IX, 340 pages. 1984.

Vol. 1042: A. Gut, K. D. Schmidt, Amarts and Set Function Processes. III, 258 pages. 1983.

Vol. 1043: Linear and Complex Analysis Problem Book. Edited by V. P. Havin, S. V. Hruščëv and N. K. Nikol'skii. XVIII, 721 pages. 1984.

Vol. 1044: E. Gekeler, Discretization Methods for Stable Initial Value Problems. VIII, 201 pages. 1984.

Vol. 1045: Differential Geometry. Proceedings, 1982. Edited by A. M. Naveira. VIII, 194 pages. 1984.

Vol. 1046: Algebraic K–Theory, Number Theory, Geometry and Analysis. Proceedings, 1982. Edited by A. Bak. IX, 464 pages. 1984.

Vol. 1047: Fluid Dynamics. Seminar, 1982. Edited by H. Beirão da Veiga. VII, 193 pages. 1984.

Vol. 1048: Kinetic Theories and the Boltzmann Equation. Seminar, 1981. Edited by C. Cercignani. VII, 248 pages. 1984.

Vol. 1049: B. Iochum, Cônes autopolaires et algèbres de Jordan. VI, 247 pages. 1984.

Vol. 1050: A. Prestel, P. Roquette, Formally p-adic Fields. V, 167 pages. 1984.

Vol. 1051: Algebraic Topology, Aarhus 1982. Proceedings. Edited by I. Madsen and B. Oliver. X, 665 pages. 1984.

Vol. 1052: Number Theory. Seminar, 1982. Edited by D. V. Chudnovsky, G. V. Chudnovsky, H. Cohn and M. B. Nathanson. V, 309 pages. 1984.

Vol. 1053: P. Hilton, Nilpotente Gruppen und nilpotente Räume. V, 221 pages. 1984.

Vol. 1054: V. Thomée, Galerkin Finite Element Methods for Parabolic Problems. VII, 237 pages. 1984.

Vol. 1055: Quantum Probability and Applications to the Quantum Theory of Irreversible Processes. Proceedings, 1982. Edited by L. Accardi, A. Frigerio and V. Gorini. VI, 411 pages. 1984.

Vol. 1056: Algebraic Geometry. Bucharest 1982. Proceedings, 1982. Edited by L. Bădescu and D. Popescu. VII, 380 pages. 1984.

Vol. 1057: Bifurcation Theory and Applications. Seminar, 1983. Edited by L. Salvadori. VII, 233 pages. 1984.

Vol. 1058: B. Aulbach, Continuous and Discrete Dynamics near Manifolds of Equilibria. IX, 142 pages. 1984.

Vol. 1059: Séminaire de Probabilités XVIII, 1982/83. Proceedings. Edité par J. Azéma et M. Yor. IV, 518 pages. 1984.

Vol. 1060: Topology. Proceedings, 1982. Edited by L. D. Faddeev and A. A. Mal'cev. VI, 389 pages. 1984.

Vol. 1061: Séminaire de Théorie du Potentiel. Paris, No. 7. Proceedings. Directeurs: M. Brelot, G. Choquet et J. Deny. Rédacteurs: F. Hirsch et G. Mokobodzki. IV, 281 pages. 1984.

Vol. 1062: J. Jost, Harmonic Maps Between Surfaces. X, 133 pages. 1984.

Vol. 1063: Orienting Polymers. Proceedings, 1983. Edited by J.L. Ericksen. VII, 166 pages. 1984.

Vol. 1064: Probability Measures on Groups VII. Proceedings, 1983. Edited by H. Heyer. X, 588 pages. 1984.

Vol. 1065: A. Cuyt, Padé Approximants for Operators: Theory and Applications. IX, 138 pages. 1984.

Vol. 1066: Numerical Analysis. Proceedings, 1983. Edited by D.F. Griffiths. XI, 275 pages. 1984.

Vol. 1067: Yasuo Okuyama, Absolute Summability of Fourier Series and Orthogonal Series. VI, 118 pages. 1984.

Vol. 1068: Number Theory, Noordwijkerhout 1983. Proceedings. Edited by H. Jager. V, 296 pages. 1984.

Vol. 1069: M. Kreck, Bordism of Diffeomorphisms and Related Topics. III, 144 pages. 1984.

Vol. 1070: Interpolation Spaces and Allied Topics in Analysis. Proceedings, 1983. Edited by M. Cwikel and J. Peetre. III, 239 pages. 1984.

Vol. 1071: Padé Approximation and its Applications, Bad Honnef 1983. Prodeedings. Edited by H. Werner and H.J. Bünger. VI, 264 pages. 1984.

Vol. 1072: F. Rothe, Global Solutions of Reaction-Diffusion Systems. V, 216 pages. 1984.

Vol. 1073: Graph Theory, Singapore 1983. Proceedings. Edited by K.M. Koh and H.P. Yap. XIII, 335 pages. 1984.

Vol. 1074: E.W. Stredulinsky, Weighted Inequalities and Degenerate Elliptic Partial Differential Equations. III, 143 pages. 1984.

Vol. 1075: H. Majima, Asymptotic Analysis for Integrable Connections with Irregular Singular Points. IX, 159 pages. 1984.

Vol. 1076: Infinite-Dimensional Systems. Proceedings, 1983. Edited by F. Kappel and W. Schappacher. VII, 278 pages. 1984.

Vol. 1077: Lie Group Representations III. Proceedings, 1982–1983. Edited by R. Herb, R. Johnson, R. Lipsman, J. Rosenberg. XI, 454 pages. 1984.

Vol. 1078: A.J.E.M. Janssen, P. van der Steen, Integration Theory. V, 224 pages. 1984.

Vol. 1079: W. Ruppert. Compact Semitopological Semigroups: An Intrinsic Theory. V, 260 pages. 1984

Vol. 1080: Probability Theory on Vector Spaces III. Proceedings, 1983. Edited by D. Szynal and A. Weron. V, 373 pages. 1984.

Vol. 1081: D. Benson, Modular Representation Theory: New Trends and Methods. XI, 231 pages. 1984.

Vol. 1082: C.-G. Schmidt, Arithmetik Abelscher Varietäten mit komplexer Multiplikation. X, 96 Seiten. 1984.

Vol. 1083: D. Bump, Automorphic Forms on GL (3,IR). XI, 184 pages. 1984.

Vol. 1084: D. Kletzing, Structure and Representations of Q-Groups. VI, 290 pages. 1984.

Vol. 1085: G.K. Immink, Asymptotics of Analytic Difference Equations. V, 134 pages. 1984.

Vol. 1086: Sensitivity of Functionals with Applications to Engineering Sciences. Proceedings, 1983. Edited by V. Komkov. V, 130 pages. 1984

Vol. 1087: W. Narkiewicz, Uniform Distribution of Sequences of Integers in Residue Classes. VIII, 125 pages. 1984.

Vol. 1088: A.V. Kakosyan, L.B. Klebanov, J.A. Melamed, Characterization of Distributions by the Method of Intensively Monotone Operators. X, 175 pages. 1984.

Vol. 1089: Measure Theory, Oberwolfach 1983. Proceedings. Edited by D. Kölzow and D. Maharam-Stone. XIII, 327 pages. 1984.

Vol. 1090: Differential Geometry of Submanifolds. Proceedings, 1984. Edited by K. Kenmotsu. VI, 132 pages. 1984.

Vol. 1091: Multifunctions and Integrands. Proceedings, 1983. Edited by G. Salinetti. V, 234 pages. 1984.

Vol. 1092: Complete Intersections. Seminar, 1983. Edited by S. Greco and R. Strano. VII, 299 pages. 1984.

Vol. 1093: A. Prestel, Lectures on Formally Real Fields. XI, 125 pages. 1984.

Vol. 1094: Analyse Complexe. Proceedings, 1983. Edité par E. Amar, R. Gay et Nguyen Thanh Van. IX, 184 pages. 1984.

Vol. 1095: Stochastic Analysis and Applications. Proceedings, 1983. Edited by A. Truman and D. Williams. V, 199 pages. 1984.

Vol. 1096: Théorie du Potentiel. Proceedings, 1983. Edité par G. Mokobodzki et D. Pinchon. IX, 601 pages. 1984.

Vol. 1097: R.M. Dudley, H. Kunita, F. Ledrappier, École d'Éte de Probabilités de Saint-Flour XII – 1982. Edité par P.L. Hennequin. X, 396 pages. 1984.

Vol. 1098: Groups – Korea 1983. Proceedings. Edited by A.C. Kim and B.H. Neumann. VII, 183 pages. 1984.

Vol. 1099: C.M. Ringel, Tame Algebras and Integral Quadratic Forms. XIII, 376 pages. 1984.

Vol. 1100: V. Ivrii, Precise Spectral Asymptotics for Elliptic Operators Acting in Fiberings over Manifolds with Boundary. V, 237 pages. 1984.

Vol. 1101: V. Cossart, J. Giraud, U. Orbanz, Resolution of Surface Singularities. VII, 132 pages. 1984.

Vol. 1102: A. Verona, Stratified Mappings – Structure and Triangulability. IX, 160 pages. 1984.

Vol. 1103: Models and Sets. Proceedings, Logic Colloquium, 1983, Part I. Edited by G.H. Müller and M.M. Richter. VIII, 484 pages. 1984.

Vol. 1104: Computation and Proof Theory. Proceedings, Logic Colloquium, 1983, Part II. Edited by M.M. Richter, E. Börger, W. Oberschelp, B. Schinzel and W. Thomas. VIII, 475 pages. 1984.

Vol. 1105: Rational Approximation and Interpolation. Proceedings, 1983. Edited by P.R. Graves-Morris, E.B. Saff and R.S. Varga. XII, 528 pages. 1984.

Vol. 1106: C.T. Chong, Techniques of Admissible Recursion Theory. IX, 214 pages. 1984.

Vol. 1107: Nonlinear Analysis and Optimization. Proceedings, 1982. Edited by C. Vinti. V, 224 pages. 1984.

Vol. 1108: Global Analysis – Studies and Applications I. Edited by Yu.G. Borisovich and Yu.E. Gliklikh. V, 301 pages. 1984.

Vol. 1109: Stochastic Aspects of Classical and Quantum Systems. Proceedings, 1983. Edited by S. Albeverio, P. Combe and M. Sirugue-Collin. IX, 227 pages. 1985.

Vol. 1110: R. Jajte, Strong Limit Theorems in Non-Commutative Probability. VI, 152 pages. 1985.

Vol. 1111: Arbeitstagung Bonn 1984. Proceedings. Edited by F. Hirzebruch, J. Schwermer and S. Suter. V, 481 pages. 1985.

Vol. 1112: Products of Conjugacy Classes in Groups. Edited by Z. Arad and M. Herzog. V, 244 pages. 1985.

Vol. 1113: P. Antosik, C. Swartz, Matrix Methods in Analysis. IV, 114 pages. 1985.

Vol. 1114: Zahlentheoretische Analysis. Seminar. Herausgegeben von E. Hlawka. V, 157 Seiten. 1985.

Vol. 1115: J. Moulin Ollagnier, Ergodic Theory and Statistical Mechanics. VI, 147 pages. 1985.

Vol. 1116: S. Stolz, Hochzusammenhängende Mannigfaltigkeiten und ihre Ränder. XXIII, 134 Seiten. 1985.